Legacy of the *Luoshu*

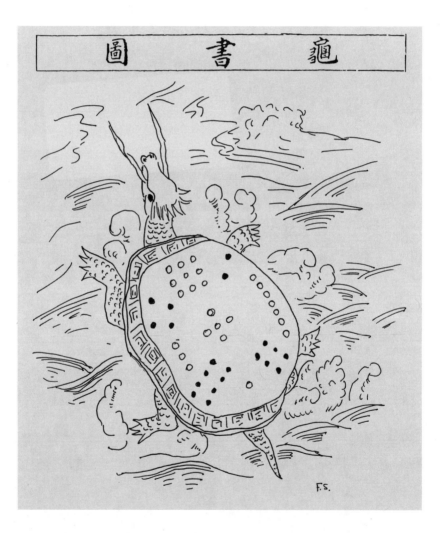

Legacy of the *Luoshu*

*The 4,000 Year Search for
the Meaning of the
Magic Square of Order Three*

FRANK J. SWETZ

OPEN COURT
Chicago and La Salle, Illinois

To order books from Open Court, call 1-800-815-2280.

Open Court Publishing Company is a division of Carus Publishing Company.

© 2002 by Carus Publishing Company

First printing 2002

All rights reserved. No part of this publication may be reproduced, stored in a retrieval system, or transmitted, in any form or by any means, electronic, mechanical, photocopying, recording, or otherwise, without the prior written permission of the publisher, Open Court Publishing Company, 315 Fifth Street, P.O. Box 300, Peru, Illinois 61354-0300.

Printed and bound in the United States of America.

Library of Congress Cataloging-in-Publication Data

Swetz, Frank.
    Legacy of the luoshu : the 4,000 year search for the meaning of the magic square of order three / Frank J. Swetz.
        p. cm.
    Includes bibliographical references and index.
    ISBN 0-8126-9448-1 (pbk. : alk. paper)
    1. Magic squares.   I. Title.

QA165 .S84 2002
511'.64—dc21

2001047449

To Joan, my yin

Contents

Acknowledgments ix
Prologue xi

1. THE JOURNEY BEGINS 1

2. ON THE BANKS OF THE LUO RIVER:
   THE CHINESE ORIGINS OF THE *LUOSHU* 9
   Mystical Beginnings on the Shell of a Tortoise 9
   The Earliest References to the *Luoshu* 12
   The Traditional Method of Constructing the *Luoshu* 16

3. *YINYANG, WUXING*, AND KEY NUMBERS IN THE *LUOSHU* 19
   "Nineness" and "Fiveness" in the *Luoshu* 20
   Yinyang: The Potential to Become 27
   Wuxing: Directions for Change 31

4. THE *LUOSHU* IN COSMIC RITUAL, FORTUNE-TELLING,
   AND *FENGSHUI* 39
   Talking to the Sky God in *Mingtang* Temples 40
   Taiyi and the Daoist Dance 48
   Divination Enters the Picture: The Eight Trigrams 53
   Cycles of Time and the Flying Star System of *Fengshui* 56
   Fortune-Telling with the *Luoshu* 59
   Variations on the *Luoshu* 61

Contents

5. **CHINESE VARIATIONS ON THE *LUOSHU* THEME** 65
   Other Magic Squares of Order Three? 65
   Higher Order Magic Squares 68
   Later Work with Magic Number Arrangements 77

6. **THE MAGIC SQUARE OF ORDER THREE IN OTHER CULTURES** 79
   Who Didn't Know about Magic Squares? 79
     Babylonia 79
     Greece 80
     Egypt 82
   Who Else Knew about Magic Squares? 82
     India 83
     Tibet 89
     Japan 90
     The Islamic World 93
   Magic Squares in Latin Europe 107
   A Mathematical Interest in Magic Squares 116

7. ***LUOSHU* MISCELLANEA** 121
   Some Mathematical Considerations 121
   Luoshu Puzzles 128
   Another Time, Another Place, Another Legend 134
   Mr. Browne's Illustrious Magic Square 135
   Feel the Rhythm 140
   Taijiquan, the *Luoshu*, and Immortality 145

8. **SOME FINAL THOUGHTS** 151
   Why Did the Magic Square Originate and Flourish in China Rather Than in the West? 152
   What Happened to the *Luoshu* as a Visible Symbol of Harmony? 155
   What is the Ultimate Significance of the *Luoshu*? 156

Epilogue   161
Notes   165
Bibliography   193
Illustration Acknowledgments   207
Index   209

Acknowledgments

In the undertaking of this study on the origins and uses of the *luoshu*, many people have assisted and advised me. I would like to acknowledge and thank them for their help and encouragement. Howard Sachs, Dean of Research and Graduate Studies at the Pennsylvania State University Harrisburg, and the Research Council supplied a modest grant that helped initiate my work on the *luoshu*. The staff of the Heindel Library at Penn State Harrisburg, particularly Ruth Runion-Slear, was most helpful in securing obscure reference material and in clarifying questionable information. Sue Hipple supplied expert and patient typing support. Several colleagues and friends came to my aid when language assistance was needed: Kirk Barbour, with French; Ali Behagi, with Farsi; Yohchia Chen, with Chinese; Refik Culpan, with Turkish and Horst Meilbrandt, with German. Fruitful research leads were provided by: Mohammad Bagheri, Tehran; J. L. Berggren, Simon Fraser University; Jean Canteins, St. Paul, France; Martin Gardner, Hendersonville, N.C.; Noël Golvers, Ferdinand Verbiest Foundation, Leuven; Takao Hayashi, Doshisla University; Ho Peng Yoke, Needham Research Institute; Karen Dee Michaelowicz, Langley School, McLean, Virginia; Yoshimasa Michiwaki, Maebashi Institute of Technology; Seyyed Hossein Nasr, George Washington University; P. Rajagapal, York University, Canada; Helaine Selin, Hampshire College; Jacques Sesiano, Ecole Polytechnique Fedérale de Lansanne; David Singmaster, South Bank University, and Robin Wilson of the Open University. David Ramsay Steele and Kerri Mommer of Open Court Publishing Company provided editorial assistance.

Such widespread interest and support reflects on the appeal and intrigue of the *luoshu* and its history, an appeal and intrigue that is still very captivating. I hope the reader will share this impression and experience the spell of magic cast by this simple number square.

Prologue

How I Got Interested in Magic Squares

In the fall of 1968 while searching the New York Public Library for materials on the history of Chinese mathematics education, I came across a listing in the card catalogue that caught my attention—"Mandarin Squares." Upon obtaining the designated book, I found that it was an account of Simon de la Loubère's experiences as envoy to the Kingdom of Siam from 1687 to1688.[1] France had established diplomatic relations with the Siamese Court in 1680. While the English language translation of de la Loubère's title page enticed the reader with the promise: "Wherein a full and curious Account is given of the Chinese Way of Arithmetick and Mathematick Learning," little substantive information was provided on Chinese mathematics or how it was taught; however, the book contained a fascinating account of Indian work with magic squares—the "Mandarin Squares" that I saw cited in the card catalogue.[2] I dismissed the book, noting its call number for future reference, but I was left with two lingering impressions: magic squares were known and used in Asia at an early date and they were somehow associated with the Chinese. These were two facts I previously had not known. Before the year was over, my research had revealed:

- the Chinese claim to the magic square of order three, also known as the *luoshu* (meaning "Luo River writing");
- its legendary origins in antiquity, possibly extending back to the third millennium B.C.E.; and
- its association with the rise of science and mathematics in China.

I sorted through this information and arrived at a tentative but relatively satisfying perspective on the nature of magic squares in early Chinese society. My major research objective was to investigate the evolution of Chinese mathematics education and magic squares, per se, were of minor concern.[3] But, as in any research project, unanswered questions remained:

- In what context was the magic square developed in China?
- How did the Chinese use their magic squares?
- Did they develop a mathematical theory of magic squares?
- Did the Chinese theories concerning magic squares influence other cultures, including those of Europe?

Eventually, I discovered the writings of Schuyler Cammann which partially resolved some of the *luoshu* issues that troubled me but they also gave rise to other issues.[4] Ultimately, I was compelled personally to investigate the mathematical and mystical impact of the *luoshu*. Now, thirty years after my initial contact with the subject, I attempt to find answers for the questions that still plague me regarding the *luoshu* and its influence on those who have come into contact with it.

The following chapters represent the sharing of my findings with a larger audience. At the onset it should be noted that I am neither a trained Sinologist nor a linguist fluent in Chinese languages but rather someone interested in the history and development of mathematical and scientific concepts and how they evolve within a society. My research approach is focused on the *luoshu*. It is selective but traverses a wide scope of concerns impinging on topics from sociology, cosmology, numerology, metaphysics, religion, mythology, mathematics, and occult beliefs. As such, nuances will remain that should be investigated further. I have provided a comprehensive bibliography to assist interested readers in this effort. This book is my attempt to answer questions, to fit pieces into a puzzle. I hope I have achieved this goal but I also hope that this work gives rise to further questions in whose pondering a better understanding of the role of mathematics in society is realized.

Among the modern Western scholars to examine Chinese rituals, customs, and traditions, perhaps Marcel Granet was the first to appreciate the deep intellectual significance surrounding the *luoshu* and its uses in Chinese society. In his considerations of Chinese ritual numerology, Granet made reference to some possible meanings associated with the *luoshu* and urged that further research on the subject be undertaken.[5] Schuyler Cammann of the University of Pennsylvania took up this task in the 1960s and firmly established that the magic square of order three was linked with early Chinese philosophy and religion. Cammann also documented the influence of the *luoshu* on Islamic and Hindu mystical thinking.[6] In turn his work inspired and stimulated other authors to investigate the relationship between magic squares and metaphysical theories. In 1973, Ho Peng Yoke published a survey of the development and evolution of magic squares in both the East and West.[7] More recently, Lars Berglund of Lund University published a comprehensive study of the *luoshu* and its relations to Chinese art and architecture.[8] Berglund was deeply affected by Cammann's theories.

It is following in this tradition, and with the conviction that there is still much to be learned about the *luoshu* and its place in human history, that this investigation now sets forth.

1

The Journey Begins

> *As you read this book, you will see that the cultural history of the magic square known as the* luoshu *is a fascinating one that touches on cosmology, mythology, philosophy, religion, occult practices, mathematics, architecture, and even music. I have uncovered some interesting aspects of the story of how the* luoshu *came to be known in different parts of the world. One of the outstanding questions concerning the* luoshu *is: How and when did it come to be known in the West?*

A Tantalizing Discovery

In the fall of 1991, I attended a conference at the University of Pittsburgh. During my free time I sought out the cultural and intellectual diversions the large city had to offer. I visited museums and art galleries. At one museum, I came across a display of rare documents—European books and manuscripts—that reflected the scientific climate of the Renaissance. Marking the end of a long hall was a round glass display case. The centerpiece, elevated within the case, was a copy of *Astronomia Europaea*, an account of Jesuit scientific accomplishments in China during the years 1669 to 1679, written by the head of the Jesuit mission, Ferdinand Verbiest (1623–1688), and published in 1687.[1] Surrounding this impressive volume were some of Verbiest's notes concerning his work and life in China. My attention was immediately drawn to one paper whose penciled diagrams seemed to involve the *luoshu*.

The *luoshu* is a very special magic square. A magic square is a square array of numbers in which the numbers in each row, each column, and each diagonal add up to the same sum—a *magic* sum. The array in figure 1.1 is a square composed of the numbers 1 through 9, in other words, the first nine counting numbers.

4	9	2
3	5	7
8	1	6

FIGURE 1.1

It is a magic square for which the magic sum is 15. The square can be further described as being of order three; that is, it has three entries in each row, column, and diagonal. Thus, the whole configuration is made up of nine numbers, (3 × 3), three squared. This particular magic square is usually referred to by its Chinese name, *luoshu* [Luo River writing], and has been revered in the East for over two thousand years.[2] It was conceived in China and transmitted westward by Arab travelers, influencing various peoples along the journey. The *luoshu* eventually found its way into medieval European astrology. In its movement from the East to the West, the diagram that originated as a cosmic chart upon which the fortunes and fate of an empire rested ended up as an occult talisman that could secure good or bad for the bearer. For many people in the contemporary world, the *luoshu* still exerts a powerful influence. In particular, as we will see in chapter 4, it forms the basis for the Asian practice of *fengshui*.[3]

The *luoshu* has often served as an emblem of harmony and, as such, represents one of the many ways traditional cultures have sought harmony through numbers, patterns, shapes, and symmetries. For example, for the Sioux of North America, who believe "the Power of the World always works in a circle, and everything tries to be round,"[4] the circle holds great significance. The Sioux symbol for life is an empty circle; the symbol for death is a full circle; and for eternity, a circle with a dot in the center. Another example is the system of number associations arrived at by the ancient Pythagoreans, in which, for instance, the number six stood for procreation. In the Ten Commandments of the Judeo-Christian tradition, the sixth commandment—"Thou shalt not commit adultery"—deals with sex. The Latin word for six was "sex," so the modern English word "sex" may really owe its origins to the number six and the number theory of the Pythagoreans. For more on issues of harmony and mathematics in traditional cultures, see the epilogue. Now let us return to the story of my discovery.

I was astounded to learn that Jesuit missionaries in China in the seventeenth century knew about and worked with the *luoshu*! Perhaps they even communicated this knowledge back to Europe! What an exciting find this was—I hoped to investigate it further at a later date, and quickly jotted down some identification data accompanying the paper: "Misc pieces dated 1668-1671-1697; *Borg. cin.* 397." When in 1996 I began my *luoshu* investigations in earnest, I resurrected the scrap of paper bearing the "museum find" data, but maddeningly I had forgotten the name of the museum I had visited and I had foolishly neglected to note the title of the display in question. So began a long and frustrating search, initially by telephone, then finally by a personal visit to Pittsburgh to retrace my tracks of six years earlier. The visit revealed that the display had been housed in the Frick Fine Arts Building at the University of Pittsburgh; however, the exact name of the exhibit remained elusive. Concomitant with my Pittsburgh-focused search to locate the tantalizing Jesuit document, I pursued a related avenue of investigation: I decided to familiarize myself with the career and work of Ferdinand Verbiest in the hope of tracing the document I had seen in Pittsburgh by starting from its human source. This effort resulted in a correspondence with Dr. Noël Golvers, a Verbiest expert.[5] Upon my description of the document, Dr. Golvers identified it immediately as BAV, *Borg. cin.* 397, a manuscript held in the Vatican Library. In my hasty effort years earlier to record the identity of my find, I had omitted the critical initials, "BAV," which stand for *"Biblioteca Apostolica Vaticana." "Borg. cin."* stands for *"Borgeani Cinesi,"* the specific collection title. Now I had it! (See figure 1.2.[6])

For various reasons, my scholarly benefactor seriously doubted that the authorship of BAV, *Borg. cin.* 397 rested with Ferdinand Verbiest.

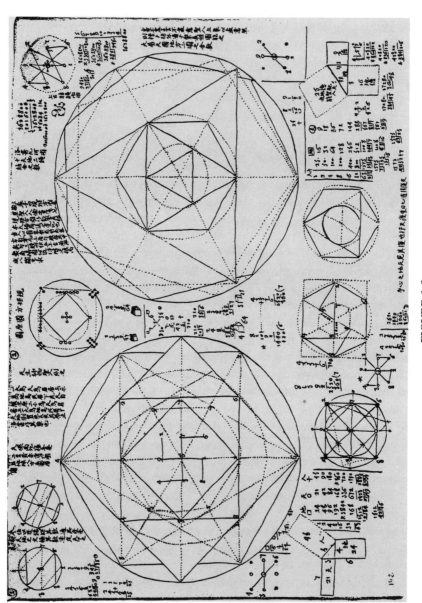

FIGURE 1.2

Rather, he was inclined to attribute it to his fellow Jesuit P. Joachim Bouvet.[7]

Setting aside the question of authorship, let me describe the contents of this seventeenth-century manuscript. The center of the group of illustrations is occupied by a stylized *luoshu* with its odd numbers (or *yang* numbers) connected by a circle, a symbol of Heaven. The even numbers (or *yin* numbers) form the vertices of a circumscribed square, a symbol of Earth. (I will discuss the connections between *yinyang* and the *luoshu* in chapter 3.) Possibly this implies that the *luoshu*, in a cosmic connotation, represents both Heaven and Earth.[8] Small diagrams at the four corners of the tableau indicate explorations of the relationships among the *luoshu*'s numerical entries: the Daoist dance path called the *yubu* is traced out in the upper right (see chapter 4 for more on Daoist dance and the *luoshu*); the sums of even and odd numbers are examined in the upper left, and the diagrams at the bottom seem to explore Pythagorean relationships based on the right triangle. Throughout the tableau, sums and products of numbers are examined for possible significant relationships. Astrological symbols are evident:) for the moon, O for the sun. One interesting computational feature is that the author designates cubed numbers, such as $4^3 = 64$ and $3^3 = 27$, by sketches of geometric cubes. But the dominating feature of the scene is the two large circles, which, in turn, circumscribe a series of regular polygons and smaller circles. In total, these two large figures embody six circles. Now if these figures are viewed as being connected sequentially—that is, with the figure on the right representing a continuation of its companion—then these six circles, each determined by the dimension of an inscribed polygon, will be concentric. At the beginning and center of this concentric series of circles lies the *luoshu*. What emerges from these geometric gyrations is a seventeenth-century European model of the universe incorporating the *luoshu*. This model confined the orbits of the planets to six circular paths determined by a nested set of celestial, crystalline spheres. In fact, the Jesuits tried in vain to convince their Chinese colleagues that celestial space was impenetrable due to the existence of confining spherical boundaries. In general, then, what the document *Borg. cin.* 397 seems to demonstrate are seventeenth-century mathematical attempts to reconcile Chinese cosmological beliefs with those prevalent in Europe at the time.

This effort by Jesuit missionaries to reconcile Chinese beliefs with Western ones coincides with a similar instance of philosophical and intellectual accommodation. The Jesuit presence in China began with

Matteo Ricci (1552–1655) who entered the Celestial Kingdom in 1583.[9] Ricci, and his later missionary successors, hoped to convert the Chinese to Catholicism and thus offset the Protestant inroads being made in Europe at the time. The Chinese, however, found the religious and mystical doctrines of Christianity to hold little appeal, but the scientific knowledge possessed by the Jesuits was another matter. These clerical envoys brought with them information on the blossoming sciences of Europe particularly: mathematics, astronomy, and metallurgy, which included cannon casting. With this knowledge, especially calendar reckoning, the Jesuits endeared themselves to the Chinese Court and secured positions in the imperial bureaucracy. Through this avenue of access to the Emperor, it was hoped that the superiority of Western ideas, including those involving religion, would become obvious to the Chinese. The Jesuits also began an active correspondence informing an eagerly awaiting European audience to the wonders of the new society they were experiencing. It was through such a correspondence that Gottfried Wilhelm Leibniz (1646–1716) became involved in the machinations of religious conversion taking place in China.

Leibniz was one of the most versatile geniuses of the seventeenth century and perhaps of all time. While primarily noted for his work in mathematics and philosophy, he was also deeply involved in studies and applications of history, law, diplomacy, politics, philology, and theology. Leibniz was a systematizer, a seeker of harmony, who attempted to organize human thought and actions through a system of logic. He was a fervent Christian who actively worked for Christian reconciliation in Europe. As an avid correspondent, Leibniz reached out to a wide audience, both spreading his ideas and increasing his knowledge of the world. His first expression of interest in China appeared in 1668 when he commented on the advanced state of Chinese medicine.[10] By 1679, he possessed some knowledge of the structure of written Chinese. Leibniz sought out information on this strange, distant land. In 1689, while visiting Rome, he met Claudio Grimaldi, a Jesuit who had spent seventeen years as a missionary in the Middle Kingdom. Grimaldi became a correspondent of Leibniz's through which the latter channeled numerous questions on Chinese life and customs. This China-centered correspondence expanded to include other China-based missionaries of the Society of Jesus among whom was the Frenchman Joachim Bouvet (1656–1730).

Both Verbiest and Bouvet had obtained a high standing in the Chinese Court and even served as scientific and mathematical tutors to the Kangxi Emperor.[11] Verbiest dismissed many of the Chinese rites and customs as hollow superstitions but Bouvet, of a more philosophical bent, saw some intellectual substance in the customs of the Middle Kingdom. In particular, Bouvet was fascinated by the theories contained in the *Yijing* [Book of Changes] and the existence of the sixty-four hexagrams which he called the "figures of Fohi" (Fuxi).[12] He envisioned that in ancient times, the Sage King Fuxi had received the divinely inspired hexagrams as a notation through which all science could be understood. Over time, the original meanings and purpose of the hexagrams had been forgotten. As a result, they had been relegated to serving merely as a set of symbols used in prognostication. Further, Bouvet recognized the correspondence between the solid and broken lines forming the hexagrams and the 1 and 0 of the theory of binary arithmetic recently proposed by Leibniz.[13]

Bouvet informed Leibniz of his findings and impressions concerning the hexagrams. Leibniz was almost overwhelmed by the implications he saw in this discovery as he associated his binary arithmetic with an "Ancient Theology" whereby the void, represented by 0, is overcome by 1, representing God and, through binary notation, all numbers (creation) would follow. At the time, he noted:

> Fohi [Fuxi], the most ancient prince and philosopher of the Chinese, had understood the origin of things from unity and nothing, i.e. his mysterious figures reveal something of an analogy to Creation, containing the binary arithmetic (and yet hinting at greater things) that I rediscovered after so many thousand of years, where all numbers are written by only two notations, 0 and 1.[14]

Through his writings, Leibniz informed the European audience of the Chinese involvement with binary numbers via the "figures of Fohi." Further, he urged Bouvet to proceed with his investigations on the original uses of the hexagrams because he believed that the conversion of the Chinese could be accomplished when the true theological principles known by Fuxi were revealed. For Leibniz, binary arithmetic was the key to converting the Chinese to Christianity.

This is an amazing episode in the history of mathematics. But with this incident in mind, these questions arise: Were the contents of BAV, *Borg. cin.* 397 intended for Leibniz? If so, what might he have made out of them? With his appreciation of Chinese philosophy and religion, and believing that both the Europeans and the Chinese originally possessed the same ancient theology, how might Gottfried Wilhelm Leibniz have interpreted the *luoshu*?[15] From existing evidence it appears that the contents of *Borg. cin.* 397 never reached Leibniz and that he remained unaware of the Chinese reverence for the magic square of order three.

Leibniz did, however, work with magic squares. In 1716, he devised a third-order magic cube and submitted it for examination to the Academy of Sciences in Paris. There, Phillippe de la Hire, a magic square enthusiast, examined it, found it correct, but could not decipher its method of construction.

2

On the Banks of the Luo River: The Chinese Origins of the *Luoshu*

According to legend, a tortoise with numbers inscribed on its shell visited Sage King Yu as he stood on the banks of the Luo River, a tributary of the Yellow River. This is how the luoshu *entered the world. The first textual reference to the* luoshu *is in the writings of Zhuang Zi (369–286 B.C.E.). Mentions of the* luoshu *are not always easy to identify because they sometimes masquerade as references to something else, such as the "Nine Halls" or "Nine Palaces."*

Mystical Beginnings on the Shell of a Tortoise

The origins of Chinese history are conceived in legends, myths, and folk tales. A variety of semi-divine cultural heroes are believed to have mastered and taught the Chinese people the arts and skills necessary for their civilization. One of these heroes, thought to have flourished in the third millennium B.C.E., was Fuxi, the inventor of fishing and hunting.[1] Later in the millennium social order was established through the efforts of the three Sage Kings: Yao, Shun, and Yu the Great (d. 2197 B.C.E.), founder of the Xia Dynasty (ca. 2000–1500 B.C.E.)[2]

In the acquisition of knowledge, both Fuxi and the Sage King Yu were subjected to strange visitations. While Fuxi was standing on the banks of the Yellow River, a "dragon-horse" confronted him. On the horse's flank was a cruciform configuration of the numbers from 1 through 10. Within its footprints, the strange beast left eight diagrams or characters composed of line segments. The number configuration has become known as the *hetu* [River diagram] and the footprint characters, the *bagua* or Eight Trigrams. Later in the Zhou dynasty (1122–255 B.C.E.), it is believed, Wenwang, the father of the founder of the dynasty, developed and extended the Eight Trigrams into the system of sixty-four hexagrams which were incorporated into the *Yijing* [Book of Changes].[3] Under the threat of a devastating flood, villagers sought out Yu, who was noted for his power over the water. Yu used his knowledge and directed the people to build canals in addition to dikes to control the waters. As we are told:

He drained off the rivers and opened up the nine outlets;
he governed their channels and directed them into the nine
courses. He opened the five lakes and settled the eastern sea.[4]

Yu tamed the waters by channeling them, whereas his father, who merely tried to hold them back through the use of dikes, failed at the task. As Yu stood on the banks of the Luo River, a tributary of the Huang He or Yellow River, a tortoise emerged from the water bearing on the underplate of its shell (plastron) an array of symbols representing numbers—it was the *luoshu*.

Both the *hetu* and the *luoshu* were magical diagrams from which, the ancient Chinese thought, an understanding of the universe and humankind's place in it could be discerned. See figure 2.1. They believed that all mathematics and science evolved from the number patterns contained in these diagrams. The flood story itself is an allegorical reference to the belief that civilization emerged from a watery chaos, a belief held by many ancient peoples.

While these legends are certainly fanciful, they lend an aura of reverence to their subjects and were most probably composed many years after both the *hetu* and *luoshu* configurations were recognized and used for significant purposes. These stories still shed some light on early Chinese culture and civilization. Both Fuxi and Yu received their special knowledge while working on or near the Yellow River. Early Chinese civilization developed along rivers and on river flood plains, particularly

The Chinese Origins of the Luoshu

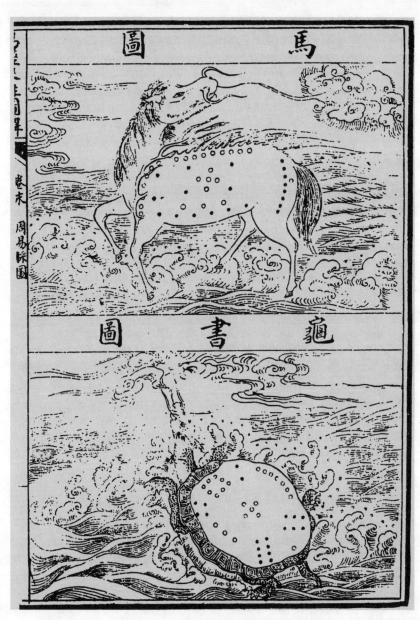

FIGURE 2.1

those of the Yangtze and Yellow Rivers. The Yellow River is so named because the large quantity of yellow loess soil in its waters makes it look yellow.[5] Annual river inundation enriches the soil and benefits agriculture but also poses hazards to riverbank dwellers. Floods must be controlled with dikes and channels. In ancient communities located near large flood-prone rivers, one of the main tasks of the ruling bureaucracy was to organize and conduct public works projects devoted to water control and conservation. Such societies have been termed "hydraulic societies."[6] China is a classic example of a hydraulic society, so it is not surprising that Fuxi's and Yu's revelations were associated with water—perhaps it could be said that their knowledge emerged from the water itself.

Together with the unicorn, the dragon, and the phoenix, the tortoise is one of the "four spiritually endowed creatures" designated in the *Book of Rites*. For the Chinese, this animal has always been a highly enigmatic and symbolic creature associated with strength, endurance, and immortality. Cultural heroes in their quests to achieve harmony or peace for the empire were often accompanied by a tortoise. The creature's shell was viewed as a model of the universe, the curved dome of the top representing the heavens and the flat, squarish underplate, the Earth.[7] Further, the plastron or underplate was partitioned along natural features into twenty-four regions correlated to the twenty-four divisions of the Chinese agricultural calendar. Ancient diviners cast such shells into fires, prognosticating fortunes from the resulting network of cracks. Later during the early Han Dynasty (ca. 200 B.C.E.), divination devices, game boards, and ceremonial ritual mirrors were modeled on the shape and pattern of the tortoise shell.[8]

The Earliest References to the *Luoshu*

The earliest references to the *luoshu* are incomplete and sporadic. This is understandable given that the use and interpretation of this number array was considered sacred, ritual practice. Knowledge of such rituals was reserved for a few select individuals, members of the priestly royal bureaucracy. Certainly, knowledge of the *luoshu* conferred on its possessor respect and esteem, social rewards highly coveted and closely guarded in Chinese society. Quite simply, those with knowledge of the *luoshu* kept it secret. Further, whatever information was actually recorded went through a "burning of the books" movement, an intellectual cleansing, ordered by the first Qin emperor, Qin Shihuangdi in 213 B.C.E.[9] Qin supposedly spared spiritual and occult writing; however, during the Sui

Dynasty (590–618 A.D.), its second emperor, Yang Di, an ardent Confucian intent on oxthodoxy, ordered the destruction of all occult books in the year 605 in order to lessen the hold of superstition on the Chinese people. Despite these obstacles of secrecy and censorship, scraps of information on the early history of the *luoshu* can still be gathered.

Initial mentions of this magic square appear during the Warring States Period, the fifth through third century B.C.E., a time of political strife and social unrest for the Chinese empire. This troubled era witnessed a flowering of philosophical speculation akin to that taking place in the Hellenic world at the same time. Wise, thoughtful people, in seeking a cure for the societal disunity around them, examined the nature of human beings and the universe as a whole as well as their relationships to each other, and sought theories and practices that would insure harmony. It was from their deliberations that the ethical, philosophical, and many of the cosmological doctrines that have shaped traditional China arose. For example, Confucianism, Mohism, and Daoism were founded during the Warring States Period.[10]

Confucianism, which stresses human relationships and ethics, in general, diverted attention from the physical world. Mohism stressed utilitarianism, advocated asceticism, and urged world understanding based on observation and logical argumentation. Daoism, a more mystical philosophy, contends that humans live in a delicate balance with nature in a vast organistic cosmos. Every being and inanimate object possesses a consciousness that functions in accordance with the consciousness of all other objects. This doctrine of balance and harmony was enunciated in the classical text, *Yijing* (though it is essentially a Confucian text). Daoism readily assimilated existing shamanistic traditions and divination beliefs and practices and maintained elaborate mythologies and pantheons. Besides serious explorations in traditional medicine and alchemy, the religion also advocated the use of magic talismans, numerology, and charms.[11] It is within the embrace of Daoist traditions that the *luoshu* eventually gained popular recognition.

The very first textual reference to the *luoshu* appears to be in the writings of Zhuang Zi (369–286 B.C.E.), one of the founders of Daoism.[12] He mentions the "nine *luo*," a phrase which has been assumed to be a succinct reference to the nine numbers of the magic square.[13] Zou Yan (305–240 B.C.E.), the patron of Chinese magicians, is believed to have manipulated the *luoshu*. In the second century B.C.E., the astronomer/mathematician Xu Yue published *Shushu jiyi* [Memoir on Some Traditions of the Mathematical Art]. In his work, Xu discusses the "Nine

Halls Calculation" and refers to the "nine palaces" which are the entries of the *luoshu*.

Later in the first century B.C.E., there appeared *Dadai liji* [Record of Rites by the Dai the Elder], a book proporting to describe ancient Chinese rites from the Zhou dynasty. In its *Mingtang* [Bright Hall] chapter, the author discusses a cosmic temple, the *Mingtang*, an architectural model and ceremonial platform embodying many of the cosmological concepts of early China.[14] His description of the *Mingtang*'s nine rooms includes a set of numbers: 2, 9, 4; 7, 5, 3; 8, 1, 6, where, it is assumed, each number was associated with a particular room, each set of numbers to be read from right to left. Thus, under the prescribed ordering the *luoshu* emerges:

4	9	2
3	5	7
8	1	6

The *luoshu* would continue to exist under the guise of the *Mingtang*'s "Nine Halls" or "Nine Palaces" diagram for many years. Records indicate that the astronomer-mathematician Zhang Heng (78–139 C.E.) used the "Nine Halls" diagram in his system of divination.

In the sixth century, the Daoist Zhen Luan, commenting on the *Shushu jiyi*, provides us with a further description of the "Nine Halls" diagram by noting that: "two and four are the shoulders, six and eight the feet, three the left and seven the right, nine the head, one the shoe and five the center."[15] Thus he anthropomorphically depicts the *luoshu*. But it is believed that Zhen borrowed this description from a much earlier work, *The Classic of the Nine Halls of the Yellow Emperor*, which has since been lost. Finally in the tenth century, Zheng Xuan (ca. 906–989), a Daoist scholar, published diagrams of both the *luoshu* and the *hetu*. See figure 2.2. He illustrated the numerical entries of each diagram by a series of white and black dots. Sometimes individual series were connected. These dots were supposed to represent knots on a cord, a method of numerical communication attributed to the ancient Chinese.[16] This pseudo-archaic mode of depicting the *luoshu* and *hetu* diagrams was intended to lend them an air of antiquity.

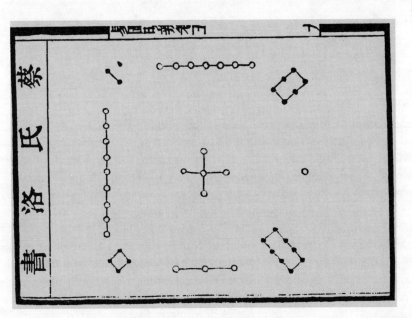

FIGURE 2.2

In the twelfth century, Cai Yuanding (1145–1198), a respected geomancer (practitioner of *fengshui*) and disciple of the eminent neo-Confucian scholar Zhu Xi, acknowledged the "Nine Halls" diagram and the *luoshu* as being one in the same. Zhu lent his support to Cai's conclusion and it became widely accepted.[17] However, by the time of the Song Dynasty (960–1279), both the *luoshu* and the *hetu* had already lost much of their cosmic significance; they survived mainly as talismans and occult diviners' devices.

The Traditional Method of Constructing the *Luoshu*

If we take the sequence of numbers from 1 to 9 and form them into a square using a lexicographical ordering, either that employed in the West, which reads from left to right, top to bottom, or the Chinese ordering that reads from top to bottom, right to left, we arrive at a "natural square" of numbers. Using the Chinese ordering, the square shown in figure 2.3 will be produced.

7	4	1
3	5	2
9	6	3

FIGURE 2.3

Examination of this array reveals that the sums of the diagonal entries and those of the second column and second row are all the same, 15. It is almost a magic square! But how to rearrange the numbers so that all rows, columns, and diagonals add up to 15? How did the Chinese do it?

The answer to this question is found in the mathematical work *Xugu zhaiqi suanfu* [Continuation of Ancient Mathematical Methods for Elucidating the Strange (Properties of Numbers)] (ca. 1275) written by the mathematician Yang Hui.[18] At that time, Yang studied the *luoshu* and several other magic squares merely as mathematical curiosities. He called the numerical configurations *zonghengtu*, literally 'transverse and longitudinal diagrams'; a term technically descriptive but devoid of magical implications. Yang begins his discussion of magic squares by demonstrating the traditional method of construction for the *luoshu*.[19] First, the natural square for the numbers 1 to 9 is constructed, then it is rotated 45 degrees counterclockwise so that the 1 is at the top and the 9 on the bottom. See figure 2.4(a). The numbers at the corners of the array are then interchanged figure 2.4(b). Now, the resulting configuration approximates Zhang Heng's association of the *luoshu* with the extremities of the human body. In fact, Yang quotes Zhang's description at this juncture of his demonstration. Finally, the numbers are compressed back into the form of a square resulting in the *luoshu*, figure 2.4(c)

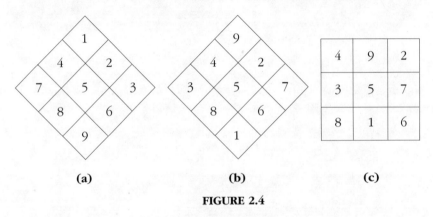

FIGURE 2.4

The *luoshu*'s legendary origins extend back into prehistory and associate it with prestigious cultural heroes. During the Zhou dynasty, it played a part in imperial rituals and in the Warring States Period, became a focus of cosmological interest. Under the guise of the "Nine Halls" diagram, it was used for consultation and divination in Han times (206 B.C.E.–220 C.E.). As a mystical symbol and charm, the *luoshu* was revered within Daoism. Given that the *luoshu* occupied a position of central importance within Chinese spiritual and metaphysical thinking for a period of over a thousand years, just what was its significance to the Chinese people? What did they see in it?

3
Yinyang, Wuxing, and Key Numbers in the *Luoshu*

> *Nine and five have special significance in the* luoshu; *nine represents the total number of cells in the* luoshu *array, and the number five occupies the important central position in the square. These same two numbers figure prominently in Chinese geography, cosmology, and metaphysics. Early Chinese writings are filled with references to groups of nine objects and many Chinese classifications are based on five.*
>
> *The "Universal Way," or Dao, expresses itself in the interaction of two primal forces that are complementary opposites—the* yin *and the* yang. *The* luoshu *lends itself to* yinyang *interpretations because odd numbers are classified as* yang *and even numbers are classified as* yin.
>
> Wuxing *theory is the theory that nature is controlled by five processes or "phases." Each phase is named for an essential material: water, fire, metal, wood, and earth. Since the numbers 1 through 9 were also associated with particular phases, the* luoshu *yields many* wuxing *interpretations.*

For the ancient Chinese, the *luoshu* represented a map of cosmic harmony, a chart for delineating interactions with the gods, heaven, and earthly institutions. It was an adaptable symbol of the Chinese world view that incorporated major forces and shapers of destiny. The interpretation of the *luoshu*'s individual numerical entries and their spatial relationships with each other were carefully correlated to reflect popular cosmological beliefs, particularly those concerning the theories of *yinyang* and *wuxing* [five phases]. In their reading of the *luoshu*, diviners and astrologers could discern the potentials for various courses of action, predict reactions, and prescribe the best path to follow. The *luoshu* offered ways to seek harmony and balance both within the realm of societal interactions and the world of nature at large.

"Nineness" and "Fiveness" in the *Luoshu*

Numerology in China has always operated at two levels: the traditional and the popular. Sometimes when they meet, these beliefs reinforce each other; at other times, they conflict. At the popular level, the magical meanings of numbers usually have been defined on the basis of homonymy and sound of the words that name them. This means that speakers of different dialects may attribute different mystical meanings to the same numbers. In Cantonese, the character for "two" is pronounced *yi* [YEE] which is also similar to the pronunciation for the characters which designate "easy" and "honor." This makes 2 an auspicious number. In contrast, the character for "four" is pronounced *si* [SEE] which sounds like the character for "death," as a consequence, 4 is considered a most unlucky number for the Cantonese and is to be avoided.[1] At the traditional level, which dominated Chinese thinking for thousands of years, numbers represent categories with particular qualities into which actions and phenomena can be conveniently sorted.[2] The practice of specifying social obligations in numbered groups thus helping to promote discipline and orthodoxy also exists in cultures outside of China: in the West, the Judaic-Christian tradition rests on observance of the "Ten Commandments"; in Islam, "Six Articles of Faith" are upheld and "Five Pillars" of obligation are pursued; and in Buddhism, the faith is built on "Four Noble Truths" and salvation or liberation is achieved by pursuing an "Eightfold Path" of proper views. But the tendency to organize concepts in numbered categories is a most striking feature of Chinese thought. For example, we encounter such groupings as: "The Gang of Four," the culprits of the Cultural Revolution (1966–1976); the "Four Wheels, the desired level of modern economic affluence (marked by the attainment of a sewing machine, a watch, and a bicycle); and the "Five Guarantees" of the Communist Party.

Two numbers, in particular, dominate the conception and operation of the *luoshu*. They are 9 and 5. In the Chinese situation, it would appear that the significance of the numbers 9 and 5 was based on mathematical considerations and then imposed on metaphysical and cosmological theories. Historically the Chinese were the first people to develop a position-based, decimal numeration system.[3] Oracle-bone inscriptions from the Yin period (ca. 14th–11th century B.C.E.) attest to this fact. By the time of the Warring States period (475–221 B.C.E.), this numeration system had been formalized in the configurations of rod numerals so named because they originated from the physical manipulation of wooden or ivory rods (small sticks) in a computational scheme. Nine primary symbols represented the numbers 1 to 9, which, of course, are the counting numerals. Zero, as a numeral, is a later historical acquisition, although its function as a placeholder was well understood and utilized at this time.[4] Thus the numbers 1 through 9 are sufficient to count all objects.[5] Nine represents completeness, fulfillment, and longevity—all desirable attributes. In old China one finds a reckoning of time based on "Nine Cycles" where each cycle represents twenty years; the imperial civil service was comprised of "Nine Grades" of Mandarins who, in turn, were rewarded for their services by a system based on "Nine Classes of Merit" that they could earn. In particular, the number 9 was associated with the emperor: royal gifts were presented in groups of nine and imperial submission was indicated by kneeling three times before the emperor and touching one's head to the ground nine times—the famous or infamous "kowtow"; the emperor's throne in the Tai Hedian Palace of the Forbidden City was adorned with nine carved dragons; and all imperial palaces had to be constructed on a podium nine feet above the surrounding surface. Grouping by nines was particularly popular in Han times for discussing geographical features. References in early Chinese literature can be found for: the "nine rivers," the "nine marshes," the "nine branches of the Yellow River," and the "nine strategic mountain locations." One description of the accomplishments of Yu the Great (he of *luoshu* tortoise fame as described above) attributed to Prince Jin of the Zhou dynasty abounds with "nines":

> [Yu the Great] stamped up high the nine mountains; dredged the nine rivers; embanked the nine marshes; flourished the nine swamps, cleared up the nine highlands, inhabited the nine far corners afar, and united the Four Seas.[6]

Even human geography (in other words, anatomy) at that time followed a similar classification system based on nine in which was found the "nine divisions of the body," the "nine viscera," and the "nine orifices."

Five was also a significant number for the ancient Chinese. The psychological basis for a decimal number system is usually attributed to the anatomical fact that humans possess ten fingers. In primitive counting situations, a one-to-one correspondence was established between an individual's fingers and the collection of objects under concern. When counting on your fingers, there is a natural break at 5 when you reach the end of one hand; thus, you really count to five and then repeat the count. The quantity 10 can be conceived of as $5 + 5$.[7] In the Chinese case, this theory is supported by the graphic configuration of their counting rod numerals which rely on the accumulation of simple tally strokes to represent the numbers up to 5 at which point a code symbol is introduced for the number 6 and carried through to 9. See figure 3.1. The numerals are written on the basis of fives: '1, 2, . . . 5, 5 and 1, 5 and 2, . . .'.[8]

FIGURE 3.1

Similarly, when the bead abacus was introduced in China during the Ming Dynasty (1368–1644) to replace the counting rod scheme, a total of 10 within a vertical column position was obtained by manipulating two, 5-valued beads. The Chinese abacus of today retains this feature and differs from other popular abacuses which use one bead to represent a 10.[9] On a mathematical note, in the sequence of numbers from 1 to 9, 5 holds the middle position. Thus, conceptually, symbolically, and operationally—the number 5 represents the center. It occupies the "middle position," designates "balance," and institutes a link that holds the other numbers in the sequence together. At times in Chinese history, the number 5 was associated with imperial power. For example, the emperor's robes bore the image of a dragon—only the emperor's dragon could have five claws, lesser dragons were limited to four claws. Within Chinese numerology 5 is a most auspicious number and many things have been classified according to five: the "Five Loves," "Five Punishments," "Five Weapons," "Five Musical Instruments," and so forth. Sinologist Wolfram Eberhard has tallied over 100 five-based categories.[10]

The "nineness" of the *luoshu* is obvious in its configuration of nine cells and nine numbers. The total sum of its entries is 45: 1 + 2 + 3 + . . . + 9 = 45; if the digits comprising 45 are then added together 4 + 5, the sum is 9. Thus one returns to or achieves completion at 9. But a more comprehensive association with "nineness" is obtained through various correlational analogies. During the Warring States period, scholars tended to divide things into groups of nine equal units of which the center unit held a position of special importance. This plan of organization may have been modeled on the concept of a rural village where eight families shared a central well. A feudal lord lived surrounded by eight vassals who benefited from his protection. Yu the Great was credited with dividing China into nine provinces, of which the central one contained the capital and the emperor. The *Shujing* [Historical Classic], attributed to Confucius, contains the *Hong Fan* or "Great Plan" chapter. This writing was divided into nine sections describing the duties and responsibilities of a sovereign. The first four sections relate how royal perfection might be achieved, the fifth and central section describes the ideal ruler and his attributes, while the last four sections focus on how to maintain royal status. In an ordering of nine, the essence of the emperor was associated with the fifth or central position. The concept of nineness and China's place in the universe was further formalized in the writings of Zou Yan (ca. 350–270 B.C.E.), a contemporary of Mencius, who theorized that China was one of nine territories comprising a continent and that nine continents comprised the world, each separated from the other by nine oceans, and from the central continent and the central territory there arose a great mountain that formed the cosmic axis around which the universe revolved.[11] China, of course, occupied the central position in this scheme—it was the "Middle Kingdom." Heaven also was believed to be divided into nine regions with the Divine Ruler living in the center. Nine mountain ranges and nine rivers were believed to interlace China and there were nine directions for purposes of orientation.

The *luoshu* could be fixed to any of these schemes. With its nine cells associated with the nine regions, halls, or palaces of Heaven, it became a celestial map and the basis of *Mingtang* ritual. Correlated with the "nine provinces," the "nine rivers," or the "nine mountain ranges," the *luoshu* became a symbolic analogue for China. More than one of these interpretations could come into play at the same time.

Figure 3.2(a) depicts the nine original provinces of China as presented in the *Shujing*. This map was made by nineteenth-century

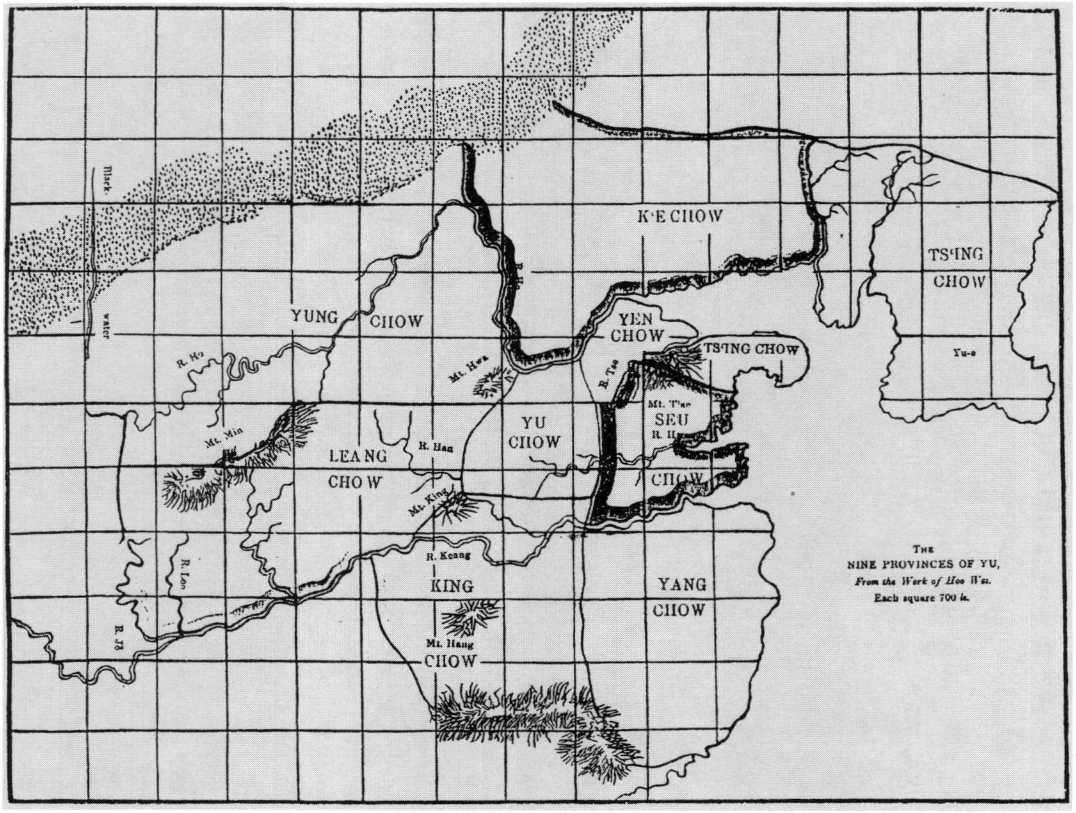

THE
NINE PROVINCES OF YU,
From the Work of Hoo Wei.
Each square 700 li.
K'E CHOW
TS'ING CHOW
Yu-e
YEN CHOW
TS'ING CHOW
YUNG CHOW
Black-water
R. Ho
Mt. Hwa
R. Tso
Mt. T'ae
SEU
CHOW
YU CHOW
Mt. Min
LEANG CHOW
R. Han
Mt. King
R. Keang
KING CHOW
Mt. Hang
YANG CHOW
R. Loo
R. Jŏ

FIGURE 3.2(a)

Yinyang, Wuxing, *and Key Numbers in the* Luoshu 25

N

Xu	Ji	Yan
Yong	Yu	Qing
Liang	Jing	Yang

W (left of square) E (right of square)

S

FIGURE 3.2(b)

Western missionaries who employed their own system of transliterating Chinese names. Figure 3.2(b) shows the corresponding *luoshu* map employing the same provinces with names given in the Pinyin system. In order to facilitate a comparison between the traditional geography of China and its symbolic *luoshu* rendering, the square is presented with the direction of north at its top. As discussed below, this was not the usual directional orientation for the *luoshu*.

The prominence of nine in the *luoshu* actually gives it a large claim. It means that the *luoshu* can be seen as the basis for all mathematics and science in China. Records of the Zhou dynasty (1027–256 B.C.E.) contain the earliest evidence of educational curricula. Children at this period were expected to study the "nine calculations of mathematics." This phrase has been interpreted in several ways but one of the most dominant understandings is that they were expected to know multiplication and division facts up to and including the nine times table. In traditional Chinese literature, the nine times table or "the table of nines" has become a synonym for mathematics.[12] Thus the *luoshu* viewed as a table of nine numbers could easily be associated with "the table of nines" and the origins of mathematics.

While the "nineness" of the *luoshu* establishes the diagram as something of a map, its "fiveness" quality provides it with a dynamic character. The number 5 occupies the key central cell. By virtue of that position, 5 figures into the calculation for the magic constant more often than any other number: four times out of eight. It is one of the addends for one column sum, one row sum, and both diagonal sums. Also, *five* times the order of the square, three, provides the magic constant, 15. This rule will work for any odd-order magic square; that is, the number in the central cell multiplied by the order of the square equals the magic sum. Similarly, the number in the central cell multiplied by the order of the square squared provides the total sum for all the elements of the square. In the case of the *luoshu*, this is 45 because: $5 \times 3^2 = 45$. But the far more interesting property of this square involving the number 5 is that 5 is the mean for each pair of outer numbers connected by a straight line through the center. That is, 5 is the mean for each of these pairs: (4, 6) (3, 7) (9, 1) (8, 2). See figure 3.3.

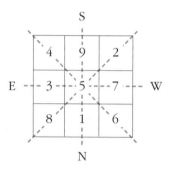

FIGURE 3.3

Thus 5 is the element that balances the square. In its pivotal position one could associate it with either China or the emperor.[13] Figure 3.3 also indicates a geographical orientation based on five principle directions that the Chinese imposed on the square: *south* at the top, *north* at the bottom, *east* on the left, *west* on the right, and the *center* in the middle.[14] By placing *north* at the bottom of maps, the Chinese felt they were "putting the cold wind" to their backs and facing the warmth of the south. In traditional China, this principle was followed in the architectural design of buildings and living compounds—the main entrance always faced south. For those of us accustomed to a system based on four principal directions, the employment of five directions may seem a bit strange, but after a little contemplation, this practice makes perfect sense.

Later commentators have pointed out that 5 is embedded in all elements of the *luoshu* and when it is subtracted, the magic square will collapse. If we subtract 5 from each element a magic square results whose magic sum is zero. See figure 3.4.

−1	4	−3
−2	0	2
3	−4	1

FIGURE 3.4

Actually what results is a magic square formed from the integers -4 to 4.

Yinyang: **The Potential to Become**

Quite early in their cosmological thinking, Chinese scholars noted the dualistic rhythms of the world around them: night followed by day; the sun succeeded by the moon in the sky; the dry season followed by the wet season; planting giving rise to harvests; and birth followed by death. In a sense, human existence took place in a realm of opposites where one state of being or status depended on another: wet, dry; hot, cold; male, female; strong, weak; active, lethargic; and so on. The contrasting qualities of each state of being and its alternate state established a potential for change, a latent energy—hot becoming cold, cold becoming hot if sources of stimuli are removed. All things thus existed in a perpetual state of flux. In their attempt to understand the conditions of change, the Chinese developed the system of *yinyang*, according to which the universe is ruled by Heaven through means of a process called the *Dao* ("the Universal Way"). Heaven acting through the *Dao* expresses itself in the interaction of two primal forces: the *yin* and the *yang*.[15] Just how this was accomplished is described more fully in the "Creation Account" given in *Huainanzi* [Book of (the Prince of) Huai-nan] (ca. 122 B.C.E.):

> Before heaven and earth had taken form all was vague and amorphous. Therefore it was called the Great Beginning. The Great Beginning produced emptiness and emptiness produced the universe. The universe produced material-force which had limits. That which was clear and light drifted up to become heaven, while that which was heavy and turbid solidified to become earth. It was very easy for the pure, fine material to come together but extremely difficult for the heavy, turbid material to solidify. Therefore heaven was completed first and earth assumed shape after. The combined essences of heaven and earth

became the *yin* and *yang,* the concentrated essences of the *yin* and *yang* became the four seasons, and the scattered essences of the four seasons became the myriad creatures of the world. After a long time the hot force of the accumulated *yang* produced fire and the essence of the fire force became the sun; the cold force of the accumulated *yin* became water and the essence of the water force became the moon. The essence of the excess force of the sun and moon became the stars and planets. Heaven received the sun, moon, and stars while earth received water and soil.

When heaven and earth were joined in emptiness and all was unwrought simplicity, then without having been created, things came into being. This was the Great Oneness. All things issued from this oneness but all became different, being divided into the various species of fish, birds, and beasts. . . . Therefore while a thing moves it is called living, and when it dies it is said to be exhausted. All are creatures. They are not the uncreated creator of things, for the creator of things is not among things. If we examine the Great Beginning of antiquity we find that man was born out of nonbeing to assume form in being. Having form, he is governed by things. But he who can return to that from which he was born and become as though formless is called a "true man." The true man is he who has never become separated from the Great Oneness.[16]

Quite simply, the *yin* and *yang* can be considered a synchronized two-force system for change. The *yang*, or male force, was the source of heat, light, and dynamic vitality and was associated with the sun, Heaven, and the seasons of Spring and Summer; in contrast, *yin*, the female force, flourished in darkness, cold, and solitude and was associated with the moon and the seasons of Autumn and Winter.[17] In conjunction, these two forces influenced all things and were present individually or together in every physical object and situation. Each force, as it reached its full strength, produced its opposite and the two continued to succeed each other in a never-ending cycle. This waxing and waning produced a perpetual state of change. While conceived in opposition the forces are not antagonistic but complementary; they depend on each other for their very existence and fulfillment. *Yin* and *yang* have been described as "the exhalation and inhalation of the universe." This view of the universe as a living, breathing entity is compatible with the idea of a universe in a constant state of flux or oscillation.

Yinyang theory provides a classification system for objects but it should be noted that their designation within the system is temporally relative and depends on the particular relationship under consideration. For example, a man relative to a woman is *yang* but a man relative to

Heaven or the gods is *yin* to the *yang* Heaven or the gods. Thus, in any *yinyang* analysis two factors must be taken into consideration: the nature of the objects in question and their relationship to each other. Initial *yinyang* designations are assigned according to some definable characteristics. For example, animals are designated *yin* or *yang* depending on how they rise up from a prone position: animals, such as horses, who rise front end first are *yang*; whereas, posterior risers, for example camels, are *yin*. In the case of numbers, odd numbers are *yang* and even numbers are *yin*.

Since the *luoshu* represented a state of harmony and cosmic equilibrium, *yin* and *yang* forces had to be balanced and had to complement each other. Zheng Xuan visually indicated this balance by representing *yin* numbers (2, 4, 6, 8) with black dots and *yang* numbers (1, 3, 5, 7, 9) with white dots. In the whole configuration there is an even number (four) of *yin* numbers and an odd number (five) of *yang* numbers. The sum of all the numbers in the diagram is 45, a number whose digits 4 and 5 represent *yin* and *yang* numbers, respectively. *Yin* and *yang* also balance the largest factors of 45, 9 and 15. When the digits of 15 are added together they give 6, a *yin* number; if the 6 is then added to *yang* 9, the sum gives the magic constant, 15.

Chinese scholars applied the *yinyang* system to directions as well. The cardinal directions plus the center, as the "strong" directions, were assigned *yang* numbers: north, 1; south, 9; the center, 5; east, 3, and west, 7; this left the weaker *yin* numbers for the subcardinal, or "weak," directions: southwest, 2; southeast, 4; northwest, 6, and northeast, 8. These orientations are shown in figure 3.5.

Southeast	South	Southwest
	4 | 9 | 2	
East	3 | 5 | 7	West
	8 | 1 | 6	
Northeast	North	Northeast

FIGURE 3.5

Still more specific orientations could be imposed on the square of numbers. Maps attributed *yin* or *yang* essences to mountains, which are strong and immovable, were considered *yang* and represented by odd numbers on maps, while rivers, sinuous, yielding, and *yin*, were represented by even numbers. During the Han Dynasty, such a scheme categorized the "five sacred mountains" of the time and the "four principal rivers": the number 5 stood for Song-Shan in Henan, Central China; 3 stood for Tai-Shan in Shandong, East China; 7 for Hua-Shan in Shaanxi, West China; 1 for Heng-Shan in Hebei, North China; and 9 for Huo-Shan in Anhui, which is considered the South Sacred Mountain. These mountains possessed a special cosmological significance. Ascending to Heaven and shrouded in mist or covered with snow, their peaks were mysterious places where the gods dwelt. They were gatherers of rain and bringers of life to the arid agricultural lands of China. To the Chinese they became conduits of *yang* energy from Heaven to Earth. As for the rivers: 4 stood for the Huai in the Southeast; 2, the San Jiang in the Southwest; 8 the Ji of the Northeast; and 6, the Yellow River found in the Northwest.[18]

Up to this point in our examination, the *luoshu* diagram has been connected with static objects and concepts in which the normally dynamic *yinyang* forces are in equilibrium and momentarily at rest. However, the theory of *yinyang* implies change and naturally lends itself to correlations with seasonal and temporal changes. Heat and light were most prominent in the summer, so this season was associated with *yang*; winter, being cold and damp, was *yin*.

Cycles of change were imposed on the *luoshu* scheme via a theory of complementary numbers. Since the Chinese conceived and employed a decimal number system, in their theory of numbers, they considered 10 the complete number in which the value of each of the first nine numbers was perfected. Thus, each number was attracted to its complement within 10: so the number 1 "moved towards" 9, 2 "moved towards" 8, and so forth. Even though odd numbers were primarily *yang* and even numbers primarily *yin* (remember the relative nature of *yinyang* designations), their complements had to possess opposite *yinyang* status—thus the complement of a *yang* number would be *yin* regardless of whether it was even or odd. The diagrams in figure 3.6 below show several mutations of the *luoshu*. In each instance the physical orientation of the parent square with *south* at the top remains the same: (a) the *luoshu* at equilibrium; (b) a *yang* cycle where *yin* elements (circled) are replaced by *yang* complements; (c) a *yin* cycle with *yang* elements (circled) replaced by *yin* complements; and (d) *yin* and *yang* at full energy level beginning decline. See figure 3.6.

The term "*yinyang*" is composed of two Chinese characters: one meaning "shade" and the other "sunshine"—the idea of "contrast" is

Yinyang, Wuxing, *and Key Numbers in the* Luoshu

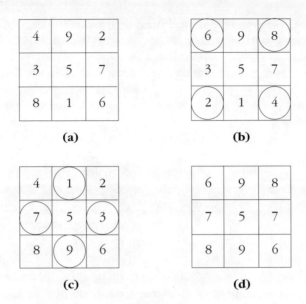

FIGURE 3.6

obvious. Its origins are attributed to Boyang Fu (ca. 8th century B.C.E.) who is believed to have associated the two metaphysical forces with earthquakes. A *yinyang* theory was further enunciated in the *Yijing* [Book of Changes] but was more fully formalized in the works of Zou Yan (ca. 350–270 B.C.E.) and he is usually credited with being the founder of the theory. *Daoism* readily absorbed *yinyang* considerations into its set of beliefs and practices, particularly those involving alchemy and medicine. Even in the present day, traditional Chinese medicine rests on the concept of *yinyang* and the maintenance of bodily harmony requires the balancing of *yinyang* forces. The rhythmic form of Chinese exercise, *taijiquan* (seen practiced in parks in China, the U.S. and throughout the world), functions on the same basis.

Wuxing: Directions for Change

The tumultuous nature of the world at large is explained in part by a passage in the *Zuo zhuan* [Master Zuo's Enlargement of the Spring and Autumn Annals] (ca. 5th century B.C.E.):

> They are Six *Qi* in nature. When they descend they give rise to the Five Tastes; display themselves in the Five Colors, and are evidenced by the Five Sounds. When they are in excess, they generate the Six

Diseases. The Six *Qi* are *yin* and *yang*, wind and rain, dark and light. They divide to form the Four Seasons, showing the Five Periods in sequence. When they are in excess, they bring about calamities. Excess in *yin* results in cold diseases; excess in *yang*, hot diseases; excess in wind, the diseases of the extremities; excess in rain, the diseases of the stomach; excess in dark, delusions; excess in light, diseases of the heart.[19]

Qi refers to primal essences from which all order in the universe proceeds. Gradually there emerged a theory that all of nature was controlled by "Five Processes" or "Five Agents," "*Wuxing*." Zou Yan stabilized the system and specified the "Five Agents": Water, Fire, Metal, Wood, and Earth. The term "*wuxing*" is difficult to translate. In the sixteenth century, Jesuit missionaries interpreted it as a theory of the "Five Elements" similar to the "Four Elements" theory held by Europeans of the time.[20] But the components of *wuxing* are not distinct substances but rather types of processes or transformations that enable change. Perhaps a better designation of the system is that of the "Five Phases."

The names of the phases are materials common and necessary in daily life: Water, Fire, Metal, Wood, and Earth. Each term does not designate the object itself but rather a type of transformation associated with the object. Water indicates energy descending, a phase in an energy cycle at which the subject has obtained a maximum point of rest. Fire shoots energy upward, under its influence energy achieves a peak. Metal indicates a dense, inward implosion of energy. Wood designates growth, an expansion of energy in all directions. Finally, Earth energy moves cyclically and horizontally about an axis. Scholars of the period envisioned the vibrations of various phenomena and correlated them with the "Five Phases," creating numerous categorical groupings of five, some of which are listed in table 3.1 below.

These are but a sampling of the categories formed according to *wuxing* theory. Some of the correlations are obvious, for example: Summer (Heat)–South –Red (the soil of southern China is red in color)–Scorched; but others, such as Fall–West–Rotten, we can only understand within the milieu of ancient China.

The principal counting numbers, 1 through 9, were also associated with the "Five Phases" according to the following scheme:

Water	Fire	Wood	Metal	Earth
1	2	3	4	5
6	7	8	9	

TABLE 3.1

	Fire	Earth	Metal	Water	Wood
Seasons	Summer	Transition between periods	Fall	Winter	Spring
Direction	South	Center	West	North	East
Color	Red	Yellow	White	Black	Green
Taste	Bitter	Sweet	Pungent	Salty	Sour
Smell	Scorched	Fragrant	Rotten	Putrid	Rancid
Creatures	Feathered	Naked	Furred	Shelled	Scaly
Sounds	Zhi	Gong	Shang	Yü	Jue[21]
Organs	Heart	Stomach	Lung	Kidney	Liver
Planets	Mars	Saturn	Venus	Mercury	Jupiter
Grains	Beans	Panicled millet	Hemp	Millet	Wheat[22]
Sense Organ	Tongue	Mouth	Nose	Ear	Eye
Domesticated Animal	Fowl	Ox	Dog	Pig	Sheep

Each pair is balanced with a *yin* number and a *yang* number; 5 retains a special status—first, because it is represented by a single number and, second, because the numbers within each pair differ by 5. Adherents of the theory of *wuxing* also categorized numbers as "heavenly" (1 to 5) and "earthly" (6 to 10). Earthly numbers were derived from heavenly numbers: 1 + 5 = 6, 2 + 5 = 7, 3 + 5 = 8, and so on. Thus, Earth, 5, is part of each season.

In a paradigm of change the "Five Phases" had to be related in some way to each other—they had to interact and yield an ordinal structure. A basic schema for change, especially in an agriculturally based society, could be modeled on the seasons as shown in figure 3.7.

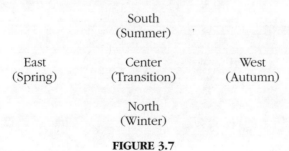

FIGURE 3.7

Zou Yan designed his *wuxing* theory primarily to explain natural phenomena; however, he also developed a system of cosmologically prescribed monarchies around the "Five Phases," associating the decline of dynasties with a particular ordering of phases.[23] This "conquest" or "destructive" cycle took place when each of the "Five Phases" conflicted with or frustrated the following phase and it is explained by the ordering: water–fire–metal–wood–earth–water. A rationale for such a conquest ordering is that water extinguishes fire, fire melts metal, metal cuts wood, wood penetrates earth, and earth (soil) absorbs water. Now, if the *luoshu*'s numbers are allowed to represent the "Five Phases," Zou's conquest cycle appears. See figure 3.8.

A counterclockwise path originating at the center, 5, spiraling downward and to the right results in the conquest cycle.

Luoshu practitioners could fit such a numbering and cycle to agricultural activities. The Chinese agricultural year began in the early spring, a period of *yang* increasing in power, symbolized by the northeast corner of the *luoshu*; the numbers 3 and 8, assigned to wood, represented growing plants. The numbers 4 and 9, in the southeast, stood for summer and were assigned to metal, perhaps in recognition of the metal tools employed in cultivation. The numbers 2 and 7, in the southwest, represented autumn and fire. Autumn may have been the time for post-harvest burning of the fields. (It should be noted, though, that the *luoshu* does not preserve the conventional associations for the directions south and west, as given in figure 3.1, when arranged as shown in figure 3.8.) Lastly, 1 and 6, in the northwest stood for winter and water, perhaps because that was the time of the flooding of fields. Earth, or soil, was central to all these activities—they revolved around it, thus

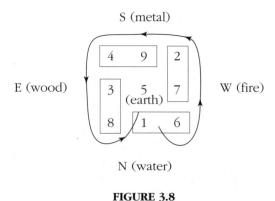

FIGURE 3.8

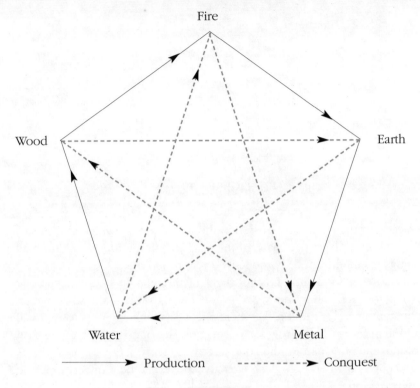

FIGURE 3.9

within the *luoshu* the cycle of agriculture, from perspective of the conquest cycle, moves counterclockwise around the number 5.

The concept of duality permeated Chinese thinking—if there was a "conquest" or "destructive" cycle, there should also exist its opposite, a "productive" or "constructive cycle." In approximately 135 B.C.E., Dong Zhongshu (fl. 179–93 B.C.E.) proposed such a cycle: wood–fire–earth–metal–water–wood: wood is nourished by water; fire is fueled by wood; earth is enriched by fire; metal is born from the contracting action of earth, and metal penetrating the earth gives rise to water. Ho Peng-Yoke has effectively illustrated the relationship of the destructive and constructive cycles in a single diagram as shown in figure 3.9.[24]

The productive cycle can also be found in the *luoshu*. If the cycle: wood–fire–earth–metal–water is traced-out within the magic square of order three allowing the central 5 to be a junction or transition point for each branch of the relationship, a swastika-like path emerges.[25] See figure 3.10. For the Chinese, this path was especially auspicious: it

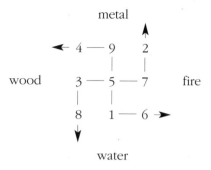

FIGURE 3.10

traced out the character "*wan*" 'ten thousand things', which also means "everything," that is, all material things. Thus, in the *luoshu* mindset, all things resulted from the productive cycle of *wuxing*.

Allowing for a little fanciful flight of thought—if one conceives of actual forces flowing from wood to fire, metal to water, and so on within the framework of the *luoshu* where 5 is a pivotal element, a torque will result and a clockwise rotation occurs, just the opposite of the direction of movement evident in the conquest cycle. The image of the *luoshu* as a "cosmic pinwheel" is indeed attractive and could be a subject of future investigation.

Visualizing the productive cycle in the *luoshu* configuration places a stress on the corners of the configuration which can be interpreted as representing the four corners of the Earth; however, in the original layout, the corner numbers represent intermediate directions and are of lesser importance than the major directional indicators. An alternate school of "Five Phases" theory sought to correct this deficiency by aligning the "Five Phases" with the five major directions. Indeed, *Luxuriant Dews of the Springs and Autumns* stresses this point in its description of the productive cycle:

> Wood produces Fire; Fire produces Earth; Earth produces Metal; Metal produces Water; and Water produces Wood. This is their father-and-son relation. Wood dwells on the left, Metal on the right, Fire in the front and Water behind, with Earth in the middle. This is also their father-and-son order, each receiving from the other in its turn.[26]

In order to accommodate this revised theory, a new cosmic diagram was devised, the *hetu*.[27] While the *hetu* came into being after the *luoshu*, legend attributed it with equally spurious ancient origins and

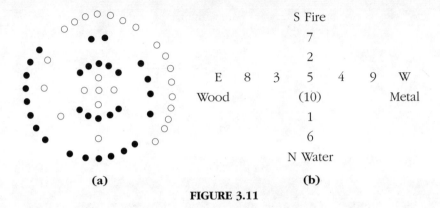

FIGURE 3.11

mysterious river and creature associations. The pseudo-archaic Daoist rendering of the diagram is given in figure 3.11(a) with a modern interpretation in figure 3.11(b).

The physical differences between the *luoshu* and *hetu* models of the "Five Phases" theory are telling. The *hetu's* cruciform configuration emphasizes "Five Directions" correlations with the opposing natures of Fire-Water and Wood-Metal visually stressed. Both the numbers 5 and 10 represent Earth, which for the old Chinese was not an uncommon practice. *Hetu* orientations interchange the South and West correlations used in the *luoshu* scheme and this revision is followed in the contemporary understanding of *wuxing* structural theory. If paths between *yang* numbers and *yin* numbers, respectively, are traced out, it will be found that they begin at 5 and spiral outwards in a clockwise direction with *yin* moving to the right of the center and *yang* to the left. The number 5 becomes the axis of the outward movement and maintains its strong central position. See figure 3.12. Excluding 5 as an addend, the

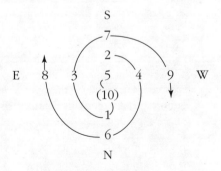

FIGURE 3.12

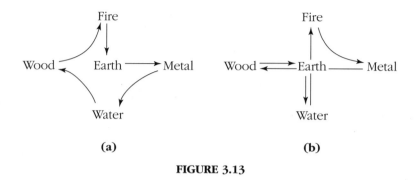

FIGURE 3.13

sum of both *yin* and *yang* numbers is equal, totaling 20—thus *yin* and *yang* are perfectly balanced.

The productive and conquest cycles are conducted along the arms of the *hetu* cross. The direction of flow for the productive cycle is clockwise whereas an examination of movement in the conquest cycle does not reveal an easily identifiable direction. See figure 3.13.

While for many, the *hetu* provided a more satisfying graphic conception of the "Five Phases" theory than its rival diagram, the *luoshu* retained its status as a convenient and approachable repository of Chinese cosmological doctrines. No doubt, the greater part of this appeal was due to the diagram's inherent mathematical properties as a magic square and its geometric shape, a bounded, finite region, easily identifiable as the Earth and as China within popular beliefs.

4

The *Luoshu* in Cosmic Ritual, Fortune-Telling, and *Fengshui*

The luoshu *was connected to the ancient cosmic Mingtang ritual which took place in temple complexes called Mingtang temples. The base of the Mingtang temple had to be square and the temple contained nine rooms, each of which came to be associated with one of the numbers from 1 to 9. The resulting number array matches the* luoshu. *In order to honor the sky god, the emperor had to follow a prescribed path in the temple, moving from one room to another each month in a certain order and moving in a certain direction. Later, Daoist priests emulated a variation of this path as part of their spiritual practice. It became the Daoist dance known as the* yubu *("steps of Yu")*

The luoshu *also became part of various other practices and devices that were intended to bring harmony or good luck or reveal information about the future. Practitioners of one ancient Chinese method of fortune-telling cast a set of rods to determine which of sixty-four possible "hexagrams" they indicate. The characteristics of the hexagrams were correlated with various cosmically significant diagrams, including the* luoshu. *The* luoshu *can also be tied to the "Flying Star" system of* fengshui. *Practitioners*

> of one method of fortune-telling in China use the numbers in a person's date of birth and place them in the luoshu grid, thus creating a natal chart that can be interpreted to provide information about the subject's personality traits, intellectual abilities, and prospects for future wealth and fortune.

Talking to the Sky God in *Mingtang* Temples

The magnificient Temple of Heaven Complex in the Beijing's southern suburbs, with its white marble Altar of Heaven and its blue-roofed Hall of Prayer for the Good Harvest, was constructed in the fifteenth century by the Yongle Emperor, the third Ming ruler, and served as a platform from which the sovereign supplicated Heaven seeking reconsecration as a ruler, absolution for the sins of his people, and the assurance of a good harvest. When it was built, the temple occupied a strategic position at the southern extreme of the city's north-south axis. Its location and alignment were chosen to attract *yang* forces. It was believed that *yin* reached its climax at the time of the winter solstice. At this time, the emperor made a pilgrimage to the Temple of Heaven and through a ritual of prayers, fasting, and sacrifice sought to attract the warm celestial force of *yang* needed to draw new crops from the kingdom's wet soil. These *yang* forces would attain their maximum influence in midsummer after which they declined. The Hall of Prayer's one entrance faced south to accommodate the arrival of the sought after *yang* energy.[1]

This annual excursion became the paramount ritual function performed by the Son of Heaven. The welfare of China and the endurance of his dynasty rested on the success, or apparent success, of the endeavor. The temple complex was a sacrosanct reserve of the emperor and his retinue—common people were forbidden to tread its grounds until 1912.[2] As instruments of communication with the gods, the Altar of Heaven and the Hall of Prayer were designed to reflect and reinforce cosmological beliefs. Both were built upon a circular, three-tiered platform whose levels stand for humankind, Earth, and Heaven. At its center, the Altar of Heaven contains a circular marble disc surrounded by concentric circles of marble flagstones distributed outward in integral multiples of nine; the first concentric circle contains nine stones, the last eighty-one. Perhaps this dependence on groupings of nine reflects the topographical numerology of Zou Yan. Zou theorized that China was composed of nine regions, for instance.[3] The blue-glazed tile roof of the Hall of Prayer symbolizes Heaven and is supported by a series of wooden columns. The four innermost columns are distributed in a square and represent the four seasons. These are surrounded by a circle of twelve columns, one for each month of the year. Lastly, an outer circle of twelve columns represents the twelve hours of the Chinese day. Thus, the temple displays features that represent both time and space—it is a cosmic temple and, as such, is the descendant of an ancient tradition extending back to the Zhou era—the *Mingtang* temple tradition.

Much has been written on the existence and use of *Mingtang* temples in ancient China; often the information given is speculative and contentious.[4] Agreement, however, can be reached on the following facts relevant to our discussion of the *luoshu*:

1. *Mingtang* temple complexes did exist. Archeological findings support this premise.[5]
2. These temples reflected the Chinese people's perception of their place in the universe and their dependence on cosmic law.
3. *Mingtang* beliefs and rituals influenced the development of an imperial political ideology based on recognizing the supreme status of the ruler.

In the earliest conceptions, *Mingtang* temples are depicted as ceremonial structures comprised of a square platform from which four pillars rose supporting a round, thatched, conical roof. The circular roof represented Heaven; the square base or platform represented Earth; and the four pillars may have represented the four seasons. In this temple, shamans made offerings to the sky gods. Over time, the architectural scheme of this structure assumed more aspects of cosmic significance, the rituals became more complex and specific, and communication with the gods became the prerogative of the ultimate mediator with Heaven, the emperor. It was in the *Mingtang* that the emperor asserted his role as the representative of humankind, the chief sacrificer to the gods. Although the idea of this temple is thought to date back to ancient times, its use flourished and was refined during the Zhou period. Later, during Han times, a revival of interest in the institution spurred a compilation and reconstitution of older writings on the *Mingtang*.[6] Some of the resulting accounts appear to have been influenced by the Han metaphysical climate. For example, Cai Yong provides these number-laden specifications:

The numerical measures of this institution all have a [cosmological] basis. The [base of the] hall is square [measuring] 144 feet [on each side], the number of the trigram *k'un* [= earth]. The roof is round with a diameter of 216 feet, the number of the trigram *ch'ien* [= heaven]. The Great Ancestral Temple of the Luminous Hall is square, measuring sixty feet [on each side], and the Chamber for Communicating with Heavens is ninety feet in diameter, [symbolizing] the changes of the yin and yang and nine and six [the numbers corresponding to the broken and unbroken lines of the hexagrams of the *Classic of Change*]. The round roof and square base [symbolize] the Way of six and nine. [The structure has] eight inner passages symbolizing the eight trigrams [of the *Classic of Change*]. It has nine rooms symbolizing the nine provinces. It has twelve palaces, thereby resonating with the twelve hours of the day. It has thirty-six doors and seventy-two windows, the number [produced by] multiplying the four doors and eight windows [of each] of the nine rooms [by nine]. The doors all open to the outside and are not shut, to illustrate that throughout the realm nothing is concealed. The Chamber for Communicating with the Heavens is eighty-one feet high, the product of the nine nines of the *huang-chung* [the tonic of the twelve pitchpipes]. [The structure's] twenty-eight pillars are evenly arrayed on the four sides, [the ones of each side] symbolizing seven of the [twenty-eight] lunar lodges. The [terrace of the] Hall is three *chang* high, thus resonating with the three calendrical cycles [used by the Hsia, Shang, and Chou Dynasties]. The five colors of its four faces symbolize [the five] phases. Its outer width is 240 feet, resonating with the twenty-four solar periods. It is surrounded on four sides by water, symbolizing the four seas.[7]

To actually build a *Mingtang* temple according to these dimensions would be impossible. Still, we have gleaned enough details of the structure from existing accounts to draw an illustration.[8] See figure 4.1.

Archeological evidence has confirmed several aspects of the historical descriptions. In particular, the square, Earth-related base did exist. It was divided into nine rooms of which the central one, the *Taishi* [Grand Chamber], served a special function as the "Chamber for Communication with Heaven." Other rooms whose location designated the principal directions of orientation for the building bore special names: the south-facing room was called the *Mingtang* [the Bright Hall] (the name also applied in general to the whole complex); at the west of center was *Zhongzhang* [Assemblage of Decor]; at the east was the *Qingyang* [Blue Solarium]; the north-facing room was the *Xuantang* [the Dark Hall]. These nine rooms also served as the "Twelve Palaces," as described below.

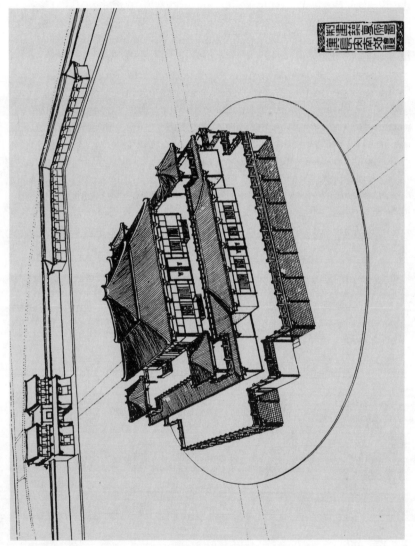

FIGURE 4.1

It was to the *Mingtang* temple that the emperor came on behalf of the people to perform the monthly tributes and sacrifices to Heaven. Such reverence was deemed necessary for harmony to prevail. The center room served as the *sanctum sanctorum* for the ceremony in which the Son of Heaven ritually communicated with the sky god. Then, in order to carry out the monthly ordinances as specified in the *Book of Rites*, the emperor had to circumambulate the complex in a clockwise direction so that he occupied a different room, or part of a room, for each month of the year in succession. Each of the rooms bore thrones facing an open window. The corner rooms of the complex contained two such thrones each. From these thrones the emperor offered his prayers for the month in the direction of the open window. The resulting twelve offertory locations comprised the "Twelve Palaces." The cycle of the royal route is shown in figure 4.2.[9]

In large part, astronomical beliefs determined the general layout of the *Mingtang* temple and the emperor's path to the twelve locations. Early Chinese astronomers believed that the North Star had a special

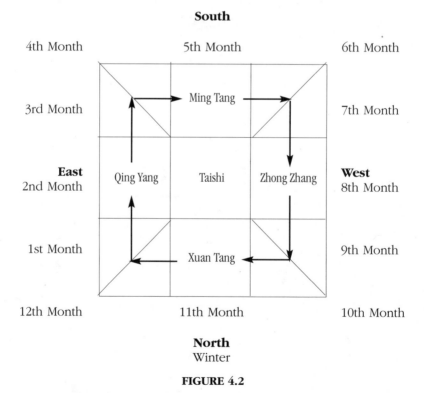

FIGURE 4.2

status. As a relatively fixed heavenly luminary around which other stars rotated in clockwise circular paths, the North Star was thought to be the abode of the sky god. Lesser heavenly gods circled in homage. They lived in eight surrounding regions of the sky. Thus, Heaven was divided into nine regions. When the supreme sky god left his purple palace at the center of Heaven to inspect his domains he traveled in a clockwise circular path. Further, an imaginary *axis mundi* originated at the Polar star and terminated in the Middle Kingdom. Thus, by analogy, the Emperor of China was associated with the supreme sky god and his actions within the *Mingtang* ceremonially paralleled those of his heavenly counterpart.

The *Book of Rites* identifies each month of the Chinese year by defining characteristics and ritual requirements. For the first month of the year it specifies:

]In the first month of Spring,
The sun is in the Ying House;
At dusk the constellation Shên culminates,
At dawn Wei culminates.

Its days are Chia and I by name;
Its Guardian is T'ai Hao,
And his attendant, Kou Mang.
Its creatures are the scaly kind;
Its note in music is Chio,
And among the standard pipes, it has the T'ai Ts'ou sound.

Its numeral is eight;
Its taste is sour;
Its smell is rank;
Its sacrifice is that of the door;
And the spleen is first offered.

The east wind sets free the frozen;
Hibernating insects begin to stir;
Fish rise up to the ice;
Otters offer up fish;
The wild geese arrive.[10]

The role of ordinary people in the seasonal drama is enacted by the emperor on the stage of the *Mingtang*.

> The Son of Heaven occupies the room to the left of the
> *Ch'ing Yang*, or Greene Bright Hall;
> He rides there in the phoenix carriage,
> Which is yoked with azure dragons wearing green flags.
> He is robed in green robes, wears green jade ornaments,
> Eats wheat with mutton;
> And his vessels are lightly carved,
> In order to aid the springing grain.[11]

> This month, we read, the Son of Heaven on the *hsin*, or first, day prays for grain from the *Shang Ti*. Then, the *yüan ch'ên* day having been selected, the Son of Heaven himself bears the plough and share, places them between the driver of his vehicle and the third occupant, the armed guard, and leads his three ducal advisers, nine ministers, the barons, and great officers, in person to plough the plot of ground of *Ti*. The Son of Heaven ploughs three furrows, the three dukes five furrows each, the ministers and barons nine each. On returning he takes hold of a goblet in the great chamber, the three dukes, nine ministers, barons, and great officers being in attendance, and gives command saying: "Wine after your Toil!"[12]
>
> After this, the text runs, the sacrificial canons are attended to, and commands are issued for sacrificing to the hills and woods, the streams and meres; but among the sacrificial victims no female victim may be used. The felling of trees is forbidden. There mustbe no disturbing of nests, and no destruction of miniature creatures, the unborn, the newly-born, or fledglings, nor of young animals, nor of eggs. There must be no great assemblages of people and no building of fortifications. Cover up skeletons and bury corpses. This month it is not permissible to take arms. To do so would entail calamities from Heaven. Military operations not-to-be-begun means that they may not commence on our side. There must be no interference with the ways of Heaven, nor any breach of the laws of earth, nor any confusion in the bonds of human Society.[13]

Dire consequences were predicted for any misapplication of the ordinances:

> If in the first month of spring the summer ordinances were carried out, then the rains would be unseasonable, plants and trees soon wither, and the state be in constant anxiety. If the autumn ordinances were carried out, then the people would suffer great epidemics, whirlwinds and violent rains would occur together, thorns, tares, weeds, and tangle flourish side by side. If the winter ordinances were carried out, then inundations would work havoc, snow and frost lay hold with strong grip, and the early seed not strike.[14]

Just as the emperor had to carry out his ritual requirements, the people of the realm were also required to perform certain acts. For example, at the time of the equinox in the second and eighth month, royal officers inspected the empire's system of weights and measures and adjusted them as needed since, as noted by Soothill, "Was not Heaven balancing night and day and the pendulum of the year? Man should take his place in the cosmic harmony." At the spring equinox, the music master was required to prepare particular instruments for the welcoming of summer.[15] Monthly ordinances closely regulated the lives of the common people by specifying what tasks they undertook, what color clothes they wore, what music and dances they performed, and even what they ate. As the sky god was the regulator of Heaven, so too, the emperor was the regulator of Earth, ordering the actions and lives of all people. From the ritual activities associated with the *Mingtang* temple the concept of the all-powerful priest-ruler-regulator developed. As Heaven was ruled from the central purple palace, so Earth, through the Son of Heaven, was ruled from the *Mingtang*. This transcendence of the emperor's influence and power was manifest in the ceremony of the "King's Assemblage."[16]

One of the important alternate functions of the *Mingtang* during the late Shang and Western Zhou periods was to serve as both a meeting site and structure for representatives of the empire to reaffirm their subservience to the Son of Heaven. In the "Grand Assemblage of People of the Four Quadrates," representatives gathered at the *Mingtang* complex and physically grouped themselves in positions symbolizing status and rank. At the center of the configuration on a raised platform, amid red tapestries and crane feathers, the emperor sat facing south. Representatives of previous dynasties, royal ministers, feudal lords, and finally barbarians were positioned around the emperor at various levels and distances, representing a hierarchy of social order and political power.[17] Even the directions the individuals faced held significance. For example, the four tribes of barbarians stood outside the four outer walls of the complex, each facing inwards towards the wall of the square and the distant emperor.

It appears that the *luoshu* was based on the layout of the *Mingtang*. The temple's square base was partitioned into nine rooms and this may have provided the cell matrix for the *luoshu*. People came to associate specific numbers with the cells, resulting in the magic square of order three. The first mention of such numbers comes from *Dadai liji* written by the Western Han scholar Dai De who apparently compiled information on the *Mingtang* from a disjointed collection of old bamboo slips. In his collage of information, there is no clear relationship between the given facts:

> *Mingtang* is an ancient institution. There are nine chambers, each
> Chamber has four doors and eight windows, altogether thirty-
> six doors and seventy-two windows. The roof is covered with
> Thatch. The top is round and the bottom square.
> *Mingtang* was so-called, because in it the rank of the lords was
> Clearly shown high or low.
> The water space [moat] outside is called *piyong*.
> The *Man* [barbarian] is in the south, *Yi* is in the east, *Di* is
> in the north and *Rong* is in the west.
> *Mingtang* is built to hold monthly observances.
> The door was Decorated red and the window white.
> The magical numbers are in the order: 2, 9, 4; 7, 5, 3; and 6, 1, 8.
> The height of the platform [foundation] is three *chi*, from the
> East to the west, it extends nine *yan* and from the south to the
> North stretches seven *yan*. The top is round and the bottom
> is square. There are nine chambers and twelve halls. Each
> chamber has four doors and each door is paired with a couple
> of windows. The palace is altogether three hundred *bu* by the
> hundred *bu* square.[18]

It was assumed that the numbers corresponded to the nine rooms of the *Mingtang* and they were assigned according to the order given. These numbers do not refer to the succession of months or to the emperor's path. They are mysterious numbers associated with particular rooms in the *Mingtang* temple. When the architectural layout of the *Mingtang* is filled in with the numbers, the resulting grid is the *luoshu*. See figure 4.3.

Taiyi and the Daoist Dance

Under the Han Emperor Wudi (140–87 B.C.E.), one god emerged from the Chinese pantheon as a deity of special importance. This was Taiyi, the Sky Emperor, who, it was thought, dwelled in a palace at the center of the night sky.[19] Further, he was envisioned as having obligations similar to his earthly counterpart, including the need to make a yearly inspection tour of his realm, the eight additional halls of his palace. Emperor Wudi made Taiyi the focus of imperial worship and even led his armies under the banner of the god. Taiyi's prominence was further

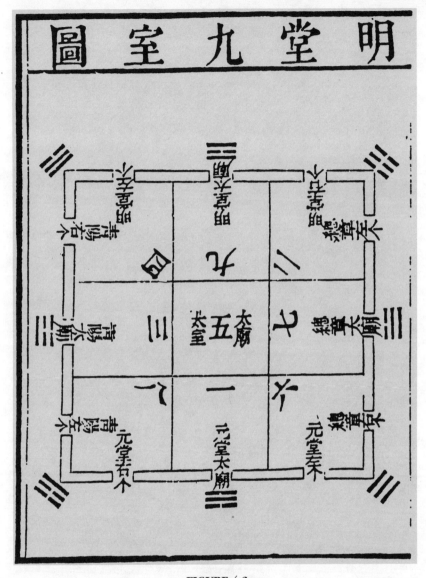

FIGURE 4.3

enhanced under the reign of the Han usurper Wang Mang (9–23 C.E.) who elevated him to the status of "Supreme Unity" and worshipped him in the *Mingtang*.[20]

Taiyi's celestial route was duplicated in ceremonies and evolved into a ritual dance for later Daoist priests. His imagined route is as follows:

he leaves his hall at the center, proceeds to the North Hall, then to the Southwest, then to the East, and then the Southeast; from here, he returns to the Center Hall, then visits the Northwest, after which he passes on to the West; finally from the West, he moves to the Northeast, then to the South Hall and returns to the center. If the "Nine Halls," *Jiu Gong*, are numbered according to the order of visitation, the *luoshu* emerges. See figure 4.4

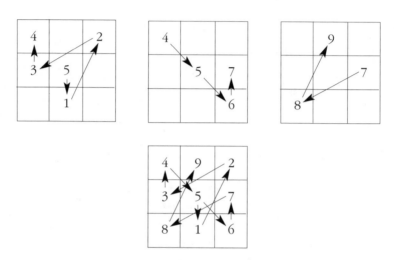

FIGURE 4.4

When committed to memory, this path becomes a kinematic algorithm for generating the magic square of order three and its associated magic squares. Daoists held the stars in the constellation known to them as *Beidou*, the Plough (also known as the Great Bear), in special reverence. While there are seven stars visible in this constellation, Daoists theorized there were actually nine.[21] In order to invoke the spiritual power of these stars, Daoist priests moved in a pattern that emulated the progression from one star to another along the path followed by Taiyi. This movement or dance became known as the *yubu*, the 'steps of Yu', in deference to the "Father of Chinese culture," the Emperor Yu. Much as Western dance manuals of the 1940s offered footprint illustrations of dance steps, Daoist manuals contained visual instructions for performing the *yubu*.[22] See figure 4.5. *Jiu Gong* (meaning the "Nine Halls") became a Daoist expression for the *luoshu*.

Medieval Daoists used the *yubu* algorithm to devise three alternate paths among the *luoshu* cells or the "Nine Halls." Together with the

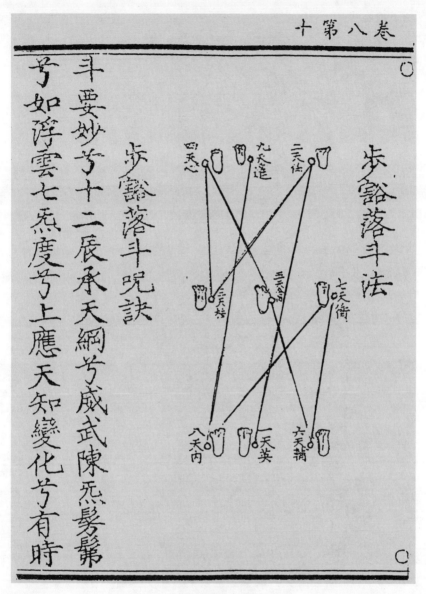

FIGURE 4.5

original *luoshu* configuration, these four paths, as they found their way into Daoist charms, became symbols for the four seasons. The alternate paths represent a series of 90-degree counterclockwise rotations of the *luoshu*. See figure 4.6.

52   *Legacy of the* Luoshu

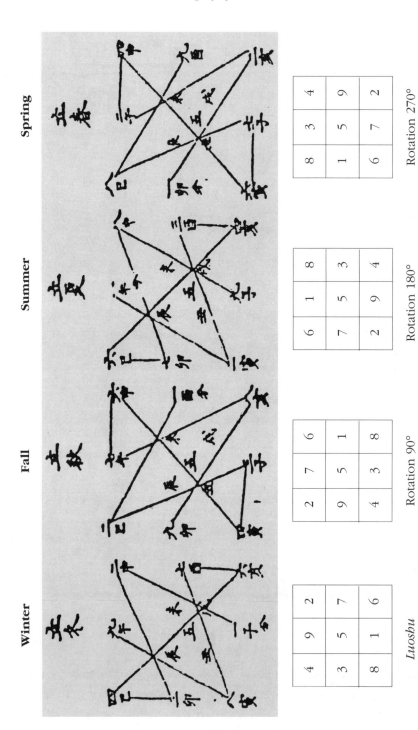

FIGURE 4.6

The Luoshu *in Cosmic Ritual, Fortune-Telling and* Fengshui   53

Illustrations of *yubu* paths were catalogued in the multivolume collection of texts that make up the Daoist canon, *Daozang* [Daoist Patrology], descriptions of beliefs and practices compiled principally during the Song Dynasty.

Divination Enters the Picture: The Eight Trigrams

One of the principal systems of divination in ancient China consisted of the casting of a set of rods, actually yarrow stalks, which resulted in two possible outcomes. Eventually, this dual possibility found meaning in the *yinyang* context and became symbolized in written characters consisting of straight line segments. A *yin* result of a rod casting was symbolized by a broken line stroke like this: – –, a *yang* result by a solid line stroke like this: —. Combining two or more rod castings created additional possibilities: the two *yinyang* results were combined to form the "Four Images"; all possible three-choice (three-stroke) combinations of *yinyang* occurrences gave rise to the *bagua*, the "Eight Trigrams," which, in turn, were combined to arrive at the sixty-four hexagrams. This progression up to the *bagua* stage is illustrated in figure 4.7.

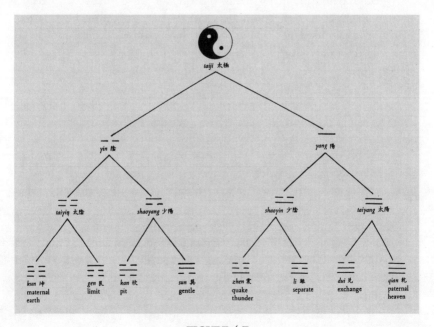

FIGURE 4.7

FIGURE 4.8

Through a rod-choosing ritual, the diviner would arrive at a hexagram.[23] Each hexagram could be read as the "sum" of six *yinyang* inputs or the combination of two trigrams. Metaphorical meanings were established for each hexagram and these meanings tabulated.[24] This tabulation of hexagram meanings became the substance of the *Yijing* [Book of Changes] and the ultimate reference for the diviner.[25] The nuances of rod selection and hexagram interpretation allowed for 4,096 mathematically possible responses to any given question. Of course, a diviner's creativity could expand this number even further.

At the cornerstone of this system of divination lay the Eight Trigrams. They were accorded honorific association with the legendary Emperor Fuxi (figure 4.8) and provided with names, attributes, images, and a familial role. See table 4.1.[26]

The Eight Trigrams were also associated with the eight directions or "Spirit Paths" of the compass. Two separate sequences exist for this, corresponding to the "Earlier Heaven Circle" and the "Later Heaven Circle."[27] This directional association then permitted further correlations with the cosmic diagrams of the *hetu* and the *luoshu* respectively. It was the *luoshu* that lent its structure for the purposes of divination as its association with the sky god Taiyi implied motion, and hence, change.

TABLE 4.1
Characteristics of the Eight Trigrams

Name	Attribute	Image	Family Relationship
Qian, the Creative	strong	Heaven	father
Kun, the Receptive	devoted, yielding	Earth	mother
Zhen, the Arousing	inciting movement	thunder	first son
Kan, the Abysmal	dangerous	water	second son
Gen, Keeping Still	resting	mountain	third son
Sun, the Gentle	penetrating	wind, wood	first daughter
Li, the Clinging	light-giving	fire	second daughter
Dui, the Joyous	joyful	lake	third daughter

Seeking harmony within change was a primary purpose of divination exercises. Correlating the *bagua* with the numbers of the *luoshu* produced additional correlations with the *wuxing* values of the numbers. Thus, a complex system of *yinyang*, *wuxing*, and *bagua* relationships were encompassed in the *luoshu* and its interpretations. See figure 4.9.

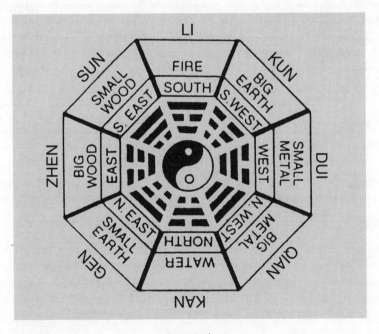

FIGURE 4.9

By calling on the numbers of the *luoshu* in some fashion, a diviner would become enmeshed in a network of possible interactions. By relying on the traditions of divination and the ultimate authority of the *Yijing*, the diviner would arrive at an advisable course of action for a client.[28] During the Song Dynasty (960–1279), examination candidates in the astronomical bureau had to be knowledgeable in three techniques of divination, two of which—*Taiyi* and *Dunjia*—relied on the *luoshu*.

Cycles of Time and the Flying Star System of *Fengshui*

Chinese astrologers devised a system of measuring time based on the *luoshu*. Among Daoists, the stars of *Beidou* were viewed as nine gods residing in nine palaces.[29] These nine deities were thought to control the destiny of humanity. When the central god moved from his palace to visit an adjoining palace, its occupant also had to move on. Thus, all the gods moved following the *yubu* path. This heavenly shuffling of positions was known as "flying across the palaces" and took nine years to complete. *Yubu* movement could be in one of two directions: clockwise or counterclockwise. *Yin* energy flows in a clockwise pattern, *yang* energy counterclockwise. In astrology, the nine phases within this cycle of movement were represented by derivatives of the *luoshu* configuration, each identified by the number and color occupying the central cell. The colors associated with the *luoshu's* elements in this system are given in figure 4.10.

4 green	9 purple	2 black
3 blue-green	5 yellow	7 red
8 white	1 white	6 white

FIGURE 4.10

The Luoshu *in Cosmic Ritual, Fortune-Telling and* Fengshui   57

Using their *zibai* [purple and white] system of identification, astrologers could specify a year of the nine-year cycle, a month within a counter-clockwise nine-month cycle, and a day within a nine-day clockwise cycle. A clockwise derivation of the nine-phase cycle is shown in figure 4.11.[30]

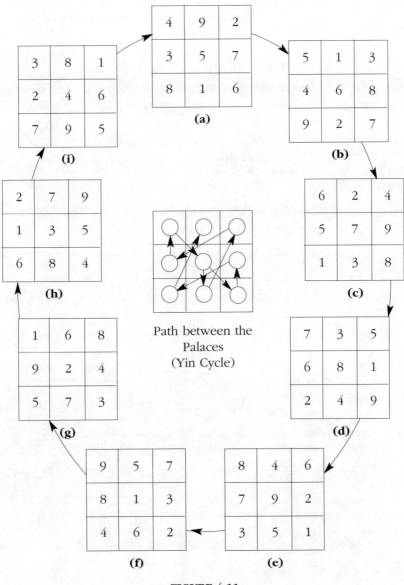

FIGURE 4.11

These phases are identified as (a) "five yellow," (b) "six white," (c) "seven red," (d) "eight white," (e) "nine purple," (f) "one white," (g) "two black," (h) "three blue-green," and (i) "four green," respectively. A counterclockwise path is generated if one starts with the *yin* complement of the *luoshu* and follows a *Yubu* path in a counterclockwise direction. In both these cyclic derivations only two of the number arrays—the *luoshu* and its *yin* complement—are truly magic squares in a mathematical sense.

A form of divination that relies heavily on the *zibai* system for its time dimension is *fengshui*, "the science of wind and water."[31] Practitioners of *fengshui* attempt to determine auspicious locations for buildings or tombs and for rooms within homes—and also the arrangement of objects within rooms—by examining the topographical features of a geographical location, using a special compass, and constructing an appropriate *zibai*-based chart. In theory, they are aligning the structure or object in question with desirable *yinyang* force fields. The computation of time in *fengshui* is based on a 180-year cycle divided into nine periods of twenty years each.[32] Individual periods are designated by a "reigning star" which corresponds with a *zibai* year. Practitioners of the "Flying Star" system of *fengshui* construct a geomantic chart for each site under consideration.[33] This chart consists of three superimposed *luoshu*-derived number squares: the first is the "Earth Square," dominated by the reigning star and arrived at by correlating the construction date with the existing 180-year cycle; secondly, the direction the front of the site faces will determine the "Facing Star" square; and finally, the direction that the rear of the site faces will determine the "Mountain Star" square. The reigning star is fixed; the earth and mountain star are movable and may, therefore, assume either *yin* or *yang* configurations. The particular orientation of these latter stars is determined by use of a geomancer's compass. When the geomantic chart is complete, each of its nine cells will contain three numbers representing *yinyang* energy levels. The diviner will interpret the interaction of these three numbers and accordingly designate the status of the site or structure and even the status of particular rooms or parts of a house.[34]

As an example of this *fengshui* technique, assume a house was built in 1972. Its reigning star is six-white (figure 4.11b). Its front is facing south, so its facing star is one-white and its mountain star is two-black. Further careful orientation and alignment indicate that the facing star is moving in the *yang* direction, the mountain star in the *yin* direction. The appropriate geomantic chart for this example is shown in figure 4.12 in

$^1 5 ^2$	$^6 1 ^6$	$^8 3 ^8$
$^9 4 ^3$	$^2 6 ^1$	$^4 8 ^8$
$^5 9 ^7$	$^7 2 ^5$	$^3 7 ^9$

FIGURE 4.12

which the normal-sized numbers represent the "Earth Square," the small superscripted numbers on the right side of each regular number belong to the "Facing Star" square, and the superscripted numbers on the left of each regular number belong to the "Mountain Star" square. Some readings of the "Earth base and Facing Star" interactions for this square indicate: children of the house will do well in academic matters (1^6); those occupying this region of the house will enjoy good health (8^8); and a threat of fire exists in this corner (9^7). Combinations of 5 and 2 are considered particularly inauspicious as they are believed to indicate sickness and accidents. Such a combination exists at the back of this house. Remedies can be undertaken: the effects of bad fortune can be reduced by placing a prescribed object in the problem room. By imposing more time constraints, that is, consultation for a special year and/or month, additional numbers can be added to the grid and more complex, detailed readings undertaken.[35] A "Flying Star" *fengshui* analysis is based primarily on time factors. More detailed *fengshui* considerations also take into account topographical factors that influence the structure.

Fortune-Telling with the *Luoshu*

In the Chinese divination arts, the magic square of order three also plays a role in predicting an individual's destiny. One popular fortune-telling method takes the numbers in a person's birth date and relates them to the numbers in the *luoshu*. According to this method, the numbers themselves, their locations, and their relationship to other numbers all influence the predictions. Within the *luoshu* grid, the uppermost horizontal row of numbers reflects on an individual's intellectual abilities, the middle row on spiritual aspects of the person's life, and the bottom row represents the material dimension. Any straight-line configuration composed of three numbers indicates a personal trait of special significance, any empty line, a special weakness. To best illustrate how this system works, let us read a fortune for an imaginary person. Assume our person is a woman born January 18, 1944. First we change this date to its Chinese lunar equivalent: 12th month, 23rd day, 1943 (12/23/43). We then place these numbers in their respective positions in the *luoshu* grid to form a natal chart (figure 4.13).

4	9	22
33		
	11	

FIGURE 4.13

According to the precepts of Chinese numerology: a 4 indicates a flair for logical thought; a single 9 indicates cleverness, ambition, and self-assurance; a double 2 is not auspicious and is associated with illness; double 3 suggests a good sense of initiative and a high degree of sensitivity; and a double 1 predicts prosperity.[36] The full row of numbers points to the existence of a powerful intellect. In total, the fortune for our imaginary individual may indicate that she is blessed with high intelligence and if she combines this intelligence with her innate understanding of people and situations it will bring her prosperity. She must, however, be on her guard to protect her health and avoid circumstances leading to illness.[37]

It is interesting to note that if a person's natal chart duplicates the *luoshu*, that is, if its nine numerical entries form the magic square of order three, then that person is considered to be perfect. Such high honor is due to the fact that the magic sum for the square would be fifteen, the number of days it takes a new moon to become full or the number of human fulfillment for the Chinese. However, due to the system of traditional dating, the maximum number of numerals possible for a birth date is eight. Human perfection, in this scheme, is impossible to achieve.

Variations on the *Luoshu*

In Chinese culture, the *luoshu* appeared in many forms contrived to suit the purposes of the moment—to reinforce a ritual, serve as a basis for prognostication, and so on. As the *luoshu* became more entrenched in Daoist beliefs, such uses became more common. Representations of the *luoshu* in the form of knotted cords or dance steps that followed the "steps of Yu" were joined by other imaginative configurations featuring caldrons, *bagua*-like lines, turtle shell anomalies, and explanatory text. See figure 4.14.[38]

Printed on paper or cloth, the *luoshu* became a powerful charm to attract spirits. On temple lanterns, the diagram symbolized a state of harmony between Heaven, Earth, and human beings and served as a map for spirits. Two such lantern motifs are given in figure 4.15.[39]

Such charms became repositories of power and were held in high esteem.

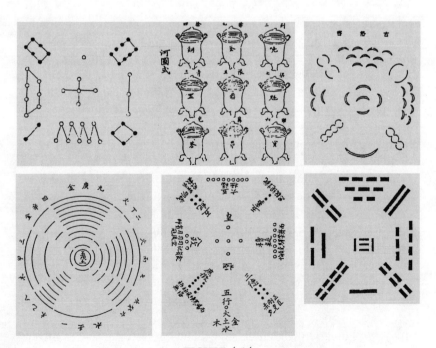

FIGURE 4.14

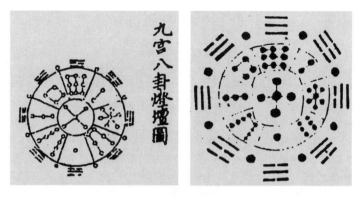

FIGURE 4.15

During the Han Dynasty, the practice of carrying a round bronze mirror to insure personal good fortune became common. The development and distribution of such mirrors reached a high point under Emperor Wang Mang who was noted for constructing a large *Mingtang* complex. The mirrors bore a highly polished surface on one side in which the user could see his or her image; the other side carried a cosmologically symbolic motif. Prominent in this motif was a square representing the Earth; at its center was a smaller square representing China.[40] Thus, in looking at this side of the mirror, one saw a square within a square within a circle—the cosmological world view of the Chinese. In addition, the larger "Earth square" bore markings indicating the four major directions, and the smaller "China square" contained a projecting knob that represented the *axis mundi* through which Heavenly power passed to Earth. In a practical sense, the knob was pierced so that a rope passed through it allowing the mirror to be attached to the belt of its beneficiary. See figure 4.16.

Inscriptions accompanying some of these mirrors urged their users to "move towards the center"—the position of power and harmony.[41] Certainly viewing one's personal image at this early date was a mystical experience in itself; in using such a mirror, the individual was psychologically and physically placing himself or herself at the center of a microcosmic universe. Thus, these "cosmic mirrors," as they came to be known, allowed their bearers to become actively involved with the metaphysical and, in a sense, undergo a mystical or religious experience. The embossed square with its symbolic accouterments represented the Earth in a cosmic context, an antecedent of the *luoshu*.[42]

FIGURE 4.16

Cosmological theories and conventions, applied on a larger scale, also influenced early Chinese architecture and the planning of cities and official compounds.[43] The association between the *luoshu* and the *Mingtang* temple has already been discussed. City planning principles were devised during the Zhou dynasty and recorded in the *Kaogong ji* [Treatise on the Examination of Artisans]. While the original copy of this text was lost, Liu Xiang of the Han Dynasty compiled a version based on then existing information. It is this text that has guided subsequent Chinese town planners up to the modern period. In the *Kaogong*, the Zhou capital was held up as a model city for emulation. It is described as being laid out in a square, nine *li* (about three miles) on a side. The sides of the city are oriented along the four major directions. Each city wall contains three gates, with the major gate being the one marking the south central position. Emanating from each gate and crossing the city is an avenue, three lanes wide to accommodate nine chariot widths; thus the city is interlaced with nine major thoroughfares. The intersection of these thoroughfares forms nine squares, of which the central one houses the royal government. A schematic plan of this ideal city is shown in figure 4.17.[44]

Architectural emphasis on a directionally oriented square, a nine-part division of the city's space, and the association of centrality with authority are all reminiscent of the *luoshu*.

Capital cities were meant to be microcosms of the universe. Most ancient Chinese capitals were destroyed and later rebuilt by succeeding dynasties. Often renovations deviated from the original plans but

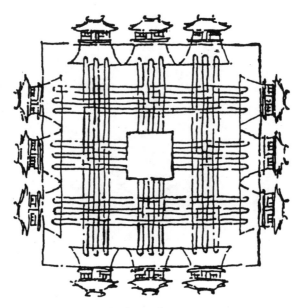

FIGURE 4.17

cosmological factors reflecting the theories of *yinyang* and *wuxing* were always followed. Every important structure was preceded by appropriate siting consultations which depended on applications of the *luoshu*.

Beijing was established as a human settlement during the Warring States era (403–221 B.C.E.) of Chinese history and served periodically as a capital for several early dynasties. In 1402, the Yongle Emperor rebuilt the city as his capital. In their construction requirements, Ming architects adhered closely to a numerology dictated by the *Yijing* and the *luoshu* stressing the "dignity of nine and five." As noted by He Junshou who made a study of this numerology: the outer city possessed seven gates, a *yang* number indicating "outer"; the inner city had nine gates, the strongest *yang* number; of the three southern gates, the central one was the most important and measured nine *zhang* in width and nine *chi* in height and the central north-south axis through the city measured fifteen *li* long—the magic constant of the *luoshu*.[45] The Forbidden City, residence of the emperor, became a microcosm within the macrocosm of Beijing and also adhered to prescribed numerology—it contained 9,999 rooms.[46]

5

Chinese Variations on the *Luoshu* Theme

> *As a result of the discussion up to this point, a natural question would seem to arise: Was the* luoshu *unique in its status as a number chart that Chinese scholars and diviners used in an attempt to understand the interplay of cosmic forces? We have seen that the* hetu *was devised as a sort of auxiliary diagram to the* luoshu *for the purposes of illustrating* wuxing *theory but that it was not a magic square. In their early proto-scientific investigations, did the Chinese employ any other magic squares? Evidence available from documents of the times would seem to indicate not. However, Yang Hui's discussion of magic squares in his 1275 work implies a surviving interest in magic squares following the* luoshu *tradition.[1] The theory of magic squares that evolved appears to have been based more on mathematical principles rather than cosmological or occult concerns.*

Other Magic Squares of Order Three?

According to Yang's technique for constructing the *luoshu*, illustrated in figure 2.4, *yang* numbers (the corner elements) were exchanged in arriving at the final results.[2] The magic square that emerges, the *luoshu*, is basically *yang*—perfectly appropriate because it represents Heaven, the ultimate source of *yang* influence. From the primary configuration given in figure 2.4(b), another magic square could be created by exchanging *yin* elements. The resulting magic square is a *yin* square and is related to the Earth. See figure 5.1.

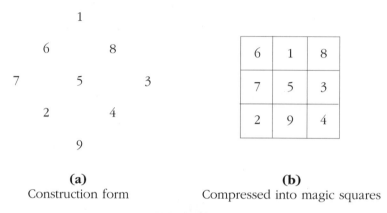

(a)
Construction form

(b)
Compressed into magic squares

FIGURE 5.1

Some evidence exists that diviners also knew of and employed this alternate square. Cammann speculates on the possibility of the two squares being used simultaneously as a model for the interplay of *yinyang* between Heaven and Earth.[3] His theory is picturesque. It was believed that one hub of the cosmic axis was located in the central room or palace of the heavenly diagram; accordingly, the other hub of the cosmic axis on Earth resided in the central cell, the "China" cell of the Earth model. The squares thus revolved around this mutually shared axis with Heaven above and Earth below. A *yang* cycle would begin with the Heaven square,

> taking every second number in its succession from the complimentary numbers in the lower square. Meanwhile, the *Yin* cycle would start in the lower, basically *Yin* square, taking its alternate numbers from the complements in the corners of the upper square.[4]

While this theory is appealing, there is no evidence that the Chinese envisioned so dramatic a ritual.

At first it might seem surprising that the Chinese did not utilize other possible variants of the *luoshu*.[5] Because it is symmetrical, the magic square of order three allows for eight permutations of its elements, producing seven other magic squares of order three that are mathematically equivalent to the *luoshu*; that is, their magic sum is 15. One can derive all of these squares from the *luoshu* configuration by physical manipulation or mathematical visualization. From the fixed position of the *luoshu*, clockwise rotations by 90 degrees about its center point or reflections about horizontal, vertical, or diagonal axes will produce equivalent magic squares. See figure 5.2.

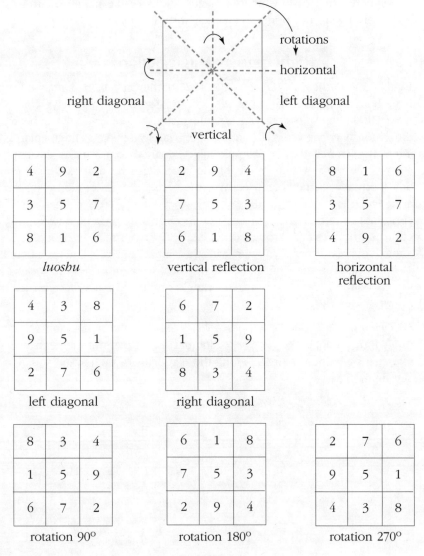

FIGURE 5.2

A rotation of 180 degrees has transformed the *luoshu yang* square into its *yin* complement. Although these permutations are easily achieved and were probably found through Chinese divination practices, the resulting squares apparently had no ritual significance. This lack of status could have been due to the restrictive nature of existing *yinyang* and *wuxing* orientations. Quite simply, the *luoshu* variants did not make cosmological sense and were thus not used. It must be remembered that early Chinese considerations of the magic square of order three were not primarily mathematical but metaphysical.

Higher Order Magic Squares

By the time of the Song Dynasty (960–1279), the Chinese had amassed a large body of knowledge on magic squares. Some of it may have come from Hindu or Arab contacts, but the majority appears to have been indigenously conceived.[6] This information was scattered and fragmentary. Yang Hui compiled findings on magic squares from various existing sources; he did not take credit for the procedures he discussed and, at times, actually seemed confused by some of them.[7] Neither did he attach a particular significance to the squares—they were merely mathematical curiosities of former times. Besides the *luoshu*, Yang presented examples of magic squares up through order ten.[8] The construction of most of Yang's squares of order n begin with the natural square of the numbers $1 - n^2$ written in the Chinese lexicographical fashion, from right to left, top to bottom. Here are some of Yang's examples that reflect on the *luoshu* tradition.

Order Four. The method employed to construct the magic square of order four parallels the *luoshu* technique in that pairs of elements are interchanged. These pairs lie along the diagonals of the natural square. See figure 5.3.

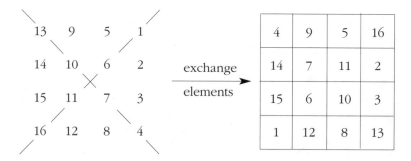

FIGURE 5.3

The resulting magic square has a magic sum of 34, a number whose digits are a balanced *yinyang* combination, 3-*yang*, 4-*yin*.

Order Five. Although Yang Hui does not provide specific construction methods for his squares beyond order four, *luoshu* techniques can be discerned. He exhibits two magic squares of order five. The first seems to be constructed by extracting a nine-number core from the natural square, figure 5.4(a); converting this core to a magic square of order three using the *luoshu yin* technique, figure 5.4(b); and then returning it to the original square and rearranging the large square's outer border elements so that opposite number pairs possess the same sum, 26, figure 5.4(c).

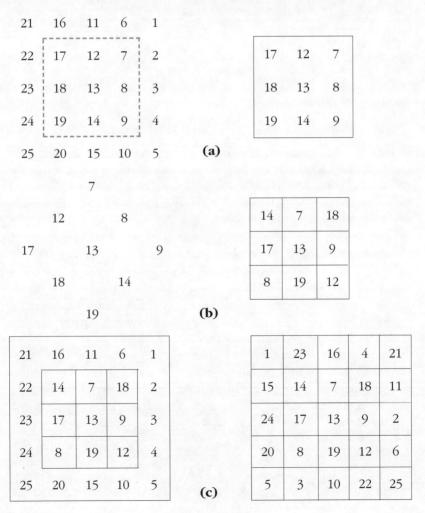

FIGURE 5.4

The second magic square of order five is more complex and is best understood by analyzing it or "tearing it down" in a reverse fashion. Its sequence of numbers runs from 9 to 33; thus it is an augmented square where 8 has been added to each entry. Subtracting 8 from each entry of the given square, its structural matrix appears. See figure 5.5.

12	27	33	23	10
28	18	13	26	20
11	15	21	17	31
22	16	29	24	14
32	19	9	15	30

subtract 8 ⟶

(4)	19	(25)	15	(2)
20	10	5	18	12
(3)	17	(13)	9	(23)
14	8	21	16	6
(24)	11	(1)	7	(22)

(a)   (b)

FIGURE 5.5

In this reduced magic square, figure 5.5(b), notice that the first four numbers, the middle number, and the last four numbers in the sequence 1 to 25 occupy relative positions in the *luoshu* configuration (the circled elements in figure 5.5[b]). Further, if 16, twice 8, is subtracted from the last four numbers of the sequence, the *luoshu* will appear. Thus, a *luoshu* framework is established. The remaining numbers in the 1 to 25 sequence are arranged in complementary pairs around the central entry, 13. Each pair totals 26. In this manner, the square is completed. Such a magic square, where elements along all diagonals through the center form complementary pairs whose sum is twice the central element, is called an "associated square." This example of a magic square of order five is both augmented and associated.

Cammann, in his study of old Chinese magic squares, offers a plausible reason as to why the ancient Chinese bothered to augment this square in the first place.[9] If the digits for the central entry of the reduced square are added together, their sum is 4; similarly, if the digits for the sum of each of the complementary pairs, 26, are added together their sum is 8, or 2×4. "Fourness," then, becomes a fundamental property of this square; 4, however, is a *yin* number, represents the Earth and is inauspicious in itself unless balanced by the *yang* number 3, which represents Heaven. By adding 8, sometimes repeatedly, to all entries of the square, the Chinese alleviated this undesirable situation.

Order Six. Historically, magic squares of order six have been difficult to devise. The Chinese used the *luoshu* principle to construct such squares of order six (shown in figure 5.6).

13	28	18	27	11	20
31	4	36	9	29	2
12	21	14	23	16	25
30	3	5	32	34	7
17	26	10	19	15	24
8	35	28	1	6	33

(a)

4	13	36	27	29	2
22	31	18	9	11	20
3	21	33	32	25	7
30	12	5	14	16	34
17	26	19	28	6	15
35	8	10	1	24	33

(b)

FIGURE 5.6

Since both of these squares were constructed by the same method, it will suffice to demonstrate the method of construction for the first square, that is, figure 5.6(a). First the numbers 1 to 36 are written in their natural sequence in four columns. Then each row of this configuration is rewritten as a 2 × 2 matrix. See figure 5.7.

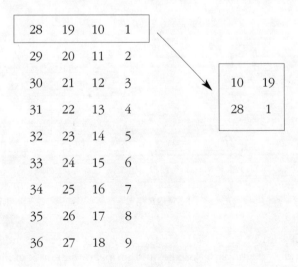

FIGURE 5.7

The resulting nine four-element number squares are then ordered according to the *luoshu* configuration with their lower right entry serving as the *luoshu* reference number. A magic square of order six results. See figure 5.8 where circled entries indicate *luoshu* reference numbers.

13	22	18	27	11	20
31	④	36	⑨	29	②
12	21	14	23	16	25
30	③	5	㉜	34	⑦
17	26	10	19	15	24
35	⑧	28	①	33	⑥

FIGURE 5.8

The set of nine four-element squares has an amazing property: if the digits for each entry are added together until a one-digit number remains, all subsquares are reduced to their *luoshu* elements, and then if the entries for each subsquare are added, a new magic square of order three appears, as shown in figure 5.9.

```
4 4      9 9      2 2
4 4      9 9      2 2

3 3      5 5      7 7
3 3      5 5      7 7

8 8      1 1      6 6
8 8      1 1      6 6
```

16	36	8
12	20	28
32	4	24

FIGURE 5.9

The total sum of each of Yang Hui's magic squares of order six is 666—a demonic symbol in medieval European numerology.[10] When these digits are added together the result is 18, and when those digits are added together the result is 9: 6 + 6 + 6 = 1 + 8 = 9, the same number obtained by adding the digits of the *luoshu*'s total sum, that is, 4 + 5 = 9. The magic square of order nine yet to be considered yields the same result.

Order Seven. Yang Hui presents two magic squares of order seven. The first is apparently constructed from the natural square for the numbers 1 to 49; however, the lexicographical ordering differs from previous instances: it goes from right to left, bottom to top. Next, a diamond-shaped core of elements is selected from the resulting number square as show in figure 5.10.

46	42	35	28	21	14	7
48	41	34	(27)	20	13	6
47	40	(33)	26	(19)	12	5
46	(39)	32	25	18	(11)	4
45	38	(31)	24	(17)	10	3
44	37	30	(23)	16	9	2
43	36	29	22	15	8	1

FIGURE 5.10

The *luoshu yin* technique is applied to the elements of this numerical array to obtain a magic square of order three, as shown in figure 5.11.

	27				23			33	23	19
33		19		33		19				
39	25		11 → 11		25		39 →	11	25	39
	31	17			31	17		31	27	17
	23				27					

FIGURE 5.11

The remaining numbers from the parent square are then arranged in complementary pairs whose sum is 50, twice the value of the central entry. This first magic square of order seven is called a "bordered square."

The second square of order seven is constructed on a *luoshu* framework. The first four numbers, the middle number, and the last four numbers of the sequence 1 to 49 occupy positions of the *luoshu* entries. Then complementary pairs of numbers whose sum is 50 are placed diagonally and equidistant around the central entry, 25. The resulting magic square is an associated square. See figure 5.12.

4	43	40	49	16	21	2
44	8	33	9	36	15	30
38	19	26	11	27	22	32
3	13	5	25	45	37	47
18	28	23	39	24	31	12
20	35	14	41	17	42	6
48	29	34	1	10	7	46

FIGURE 5.12

Order Nine. The magic square of order nine is the most complex and fascinating of all the old Chinese magic squares. Its construction seems strikingly simple. After the natural square for the numbers 1 to 81 is constructed, each row of the configuration is then folded into a 3 × 3 magic square whose central, lower, entry is a *luoshu* number. This process is illustrated in figure 5.13 using the fourth row of the natural square.

FIGURE 5.13

Chinese Variations on the Luoshu Theme

The resulting nine magic squares of order three are compiled into a square where their position is indexed according to their *luoshu* number (circled in figure 5.14).

31	76	13	36	81	18	29	74	11
22	40	58	27	45	63	20	38	56
67	(4)	49	72	(9)	54	65	(2)	47
30	75	12	32	77	14	34	79	16
21	39	57	23	41	59	25	43	61
66	(3)	48	68	(5)	50	70	(7)	52
35	80	17	28	73	10	33	78	15
26	44	62	19	37	55	24	42	60
71	(8)	53	64	(1)	46	69	(6)	51

FIGURE 5.14

A magic square composed of magic squares, such as the one in figure 5.14, is called a composite magic square. Yang Hui's composite magic square of order nine has a magic constant of 369, that is, 41×9. If each of the subsquares is replaced by their respective magic constant, a third-order magic square appears that has the same magic constant as its parent (figure 5.15a). Similarly, if each subsquare is replaced by its total sum, a third-order magic square appears whose total sum is 3,321, the same as its parent (figure 5.15b).

120	135	114
117	123	129
132	111	126

(a)

360	405	342
351	369	387
396	333	378

(b)

FIGURE 5.15

The magic square shown in figure 5.14 gives rise to another magic square of order nine with similarly interesting properties. A set of nine magic squares of the third order can be constructed by choosing all numbers that occupy the same relative position within their respective subsquare, that is, first element, first row, and so on and then forming each set of nine such numbers into a third-order magic square. In turn, each of these nine resulting subsquares will be compiled into a large square of order nine with the subsquare comprised of first element, first-row elements becoming the first element, first-row for the intended square, second element, first-row elements becoming the second element, first row, and so on. The resulting composite square arrived at is shown in figure 5.16.

31	36	29	76	81	74	13	18	11
30	32	34	75	77	79	12	14	16
35	28	33	80	73	78	17	10	15
22	27	20	40	45	38	58	63	56
21	23	25	39	41	43	57	59	61
26	19	24	44	37	42	62	55	60
67	72	65	4	9	2	49	54	47
66	68	70	3	5	7	48	50	52
71	64	69	8	1	6	53	46	51

FIGURE 5.16

Further, if the magic constant for each of these nine subsquares is computed and these results are then ordered according to size, it will be found that they correspond exactly with the value of the *luoshu* position the subsquare occupies in the large square.[11]

Yang's section on magic squares in the *Xugu* is followed by a section on magic circles and magic circle configurations. These numerical figures are ingeniously conceived and attest to the imagination and ability of their inventors; however, they do not reflect on the *luoshu*. Rather they appear to be challenging extensions of the magic square concept.[12]

Later Work with Magic Number Arrangements

While some of the *luoshu* techniques were followed by succeeding mathematicians, they were performed with little ritual or metaphysical understanding, having become merely "a means to an end"—the construction of a magic square.[13] In 1593, Chen Dawei published a series of fourteen diagrams on magic squares and circles.[14] He repeated much of Yang Hui's work. A new dimension to magic configurations was added by Fang Zhongtong (1633–1698) who worked with magic cubes and spheres as well as magic squares. In the same century, Zhang Chao (b. 1650) published China's first magic square of order ten.[15] Finally, in the late nineteenth century, Bao Qishou, a gifted amateur mathematician, wrote a book on magic configurations including three-dimensional magic number arrangements.[16]

6

The Magic Square of Order Three in Other Cultures

> *While most evidence would seem to substantiate Chinese claims that they knew of and used the magic square of order three as early as the first century C.E., these questions remain: Were the ancient Chinese really the first to discover this mathematical concept? Did other ancient peoples conceive of the idea of a magic square? Let us examine the mystical number trends and practices followed in other ancient societies.*

Who Didn't Know about Magic Squares?

Babylonia

The roots of much of Western numerology and astrology can be traced back to the peoples of the ancient Tigris and Euphrates River basins, the region which is popularly known in its historic context as Babylonia. The Babylonians believed in "lucky" and "unlucky" numbers. Thirteen was considered an unlucky number possibly because it came after twelve, an integral factor of sixty, the base for the Babylonian number system and, in itself, a poor factor within the system. Seven was a powerful number, its influence apparent in the creation myth, *Enuma Elish*:

the seven winds; the seven spirits of the storm; the seven evil diseases; and the seven divisions of the underworld contained by seven doors bearing seven seals. This reverence for the number seven passed initially into Hebrew beliefs, a little later into those of Islam, and eventually into the designations of Christian arithmology.[1] The Babylonians are the first known people to assign numerical values to words, particularly personal names. They then manipulated the numerical values to establish mystical relationships between people and/or objects. Dimensions of temples and other important structures usually bore a symbolic significance.[2] This system of numerical encoding of words was known as *isopsephia*. Later the Greeks and Romans used the system, Hebrew mystics adopted it in Kabbala, and it emerged in medieval Europe under the practices of the *gematria* system of numerology.[3] Another mystical use of numbers in old Babylonia was to establish a hierarchy for the pantheon of gods. Anu, the supreme lord, possessed the perfect number, sixty; other gods were assigned numbers of a lesser order according to their importance. In this scheme, the goddess Ishtar, lover and sharer of power with Anu, was given the number fifteen and associated with the planet Venus. The goddess of love and war, Ishtar, was often depicted with eight rays emanating from her body. Since Ishtar plus her eight powerful rays equals nine, and Ishtar embodies the number fifteen, Frans Endres, in his work on the mystical use of numbers, speculates that the Babylonians devised the *luoshu*; however, this conjecture lacks further substantiation.[4] In general, there is little evidence that the Babylonians knew of or used magic squares, despite their great interest in numerology and number manipulation.

Greece

Greek number mysticism centers around the beliefs and practices of Pythagoras of Samos (ca. 540 B.C.E.) and his followers. The Pythagoreans believed, "All is ruled by number"—that numerical relationships controlled the harmonious functioning of the universe. Number and proportion were the overriding factors in their philosophy. The number one, the "Great Monad," was not a number in itself but rather the creator of all numbers and all things. It served as the creator and also represented "reason." While the Pythagoreans assigned specific meanings to all numbers, they also grouped numbers into categories of special significance. Odd numbers were considered male numbers, even numbers, female. Major Greek deities were given an odd number, lesser deities an even number. This association of odd numbers with the divine has lingered across cultures; for example, an Islamic saying declares, "God loves the odd," Falstaff advises, "There is divinity in odd numbers" in Shakespeare's *Merry Wives of Windsor*.[5] The interactions of numbers were also given meaning; for example, the number five, as the coming together of the first female number, two, and the first male number, three, symbolized marriage for the Pythagoreans. However, Greek numerological manipulation of numbers, while quite broad and complex, did not encompass the use of magic squares. The closest the Pythagoreans came to arriving at a geo-arithmetic configuration that approximated the *luoshu* in several symbolic respects was in their concern with the *tetractys* or "holy fourness."

The *tetractys*, a sacred symbol and the conceptual basis of the Pythagorean oath, consisted of the union of the first four numbers, namely, 1 + 2 + 3 + 4, graphically represented as a configuration of dots. See figure 6.1

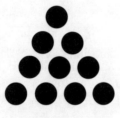

FIGURE 6.1

Pythagorean cosmology conceived of four elements, the physical building blocks of the universe; they were Fire, Earth, Air, and Water with numerical values of one, two, three, and four, respectively. The numerical value of the universe was ten, the number of completion, because the sum of the numbers of the four elements equaled 10: 1 + 2 + 3 + 4 = 10. As noted in a prayer attributed to the Pythagoreans:

> Bless us, divine number, thou who generatest [sic]
> gods and men! O holy, holy *tetractys*, thou that containest
> the root and the source of the eternally flowing creation!
> For the divine number begins with the profound, pure
> unity until it comes to the holy four; then it begets the
> mother of all, the all-comprising, the all-bounding, the
> first born, the never-serving, the never-tiring holy ten, the
> keyholder of all.[6]

Thus ten, as expressed in the Pythagorean form of a triangular number, held special mystical significance for its beholder.[7] The shape, an equilateral triangle, indicated perfection and the ratios of its number lines—2:1, 4:3, and 3:2—designated the harmonies on a musical scale of an octave, a fourth, and a fifth, respectively. The *tetractys* was a Greek expression of creation, symmetry, and harmony in the cosmos in a manner similar to that perceived in the Chinese *luoshu*, but it was not a magic square.

Chinese and Greek metaphysical theories and methodologies share certain features: (a dualistic view of nature; a primal element theory—in the Chinese case, five elements, and for the Greeks, four elements; and a psychological reliance on a geo-arithmetic diagram to express cosmic order). These similarities have led historians of mathematics to speculate that ideas migrated from Greece to China (or vice-versa).[8] The issue, however, remains unresolved.

Egypt

Little is known about ancient Egyptian number lore.[9] The Egyptians followed many of the numerological practices of their neighbors—they had "lucky" and "unlucky" numbers and participated in *isopsephia*. Mathematical knowledge was usually limited to priests and scribes, and mathematics was considered a mysterious entity in itself. The preface to the mathematical Rhind papyrus promises that its text offers a "Complete and thorough study of all things, insights into all that exists, knowledge of all secrets. . .".[10] The three-four-five right triangle was known in ancient Egypt and believed to be a symbol of "universal nature." The trinity of major gods was represented by the sides this triangle: the base of length four symbolized Osiris; the remaining leg, length three, stood for Isis, and the hypotenuse represented Horus, the son of Osiris and Isis. No knowledge of magic squares is apparent in existing Egyptian mathematical material.

Thus, among the major cultures that are usually credited with having influenced the development of Western mathematics, no evidence indicates a knowledge of or use of magic squares prior to the appearance of the Chinese *luoshu*.

Who Else Knew about Magic Squares?

Well before the Christian Era, China had established trade contacts with its neighbors and distant empires. Gradually, this intercourse expanded to include political, cultural, and scientific exchanges. The Middle Kingdom imported Buddhism from its neighbor, India. Buddhist and, eventually, Islamic scholars took jobs in the Imperial bureaucracy. Certainly, they brought new knowledge with them into China, and when they left, they were enriched by knowledge of Chinese practices and customs. How extensive this early exchange of information between cultures was remains unclear but surely esoteric theories and occult practices made appealing intellectual souvenirs to bring home from China. Let us examine the history of magic squares in cultures proximate to the Chinese.

India

Information on ancient Indian mathematics is fragmentary, and its conclusions are often contentious; however, there is sufficient evidence to indicate that a tradition of using magic squares dates back to very early times. Legend proposes that knowledge of magic squares was communicated from Lord Siva to a magician named Manibhadra. Magic squares were viewed as magical devices capable of providing supernatural assistance to their patrons. This view persists even today. The first documented mention of magic squares in India is found in Varahamihira's work on divination, *Brhatsamhita* (ca. 550 C.E.). In this text, Varahamihira uses a number square of order four to prescribe combinations and quantities of ingredients for perfume manufacture.[11] The square is comprised of the numbers 1 to 8, each employed twice within the array, thus violating the strict definition of uniqueness among elements of a magic square. Still, the configuration is "magical" in that the constant sum is 18 and that it is pan-diagonal, that is, the diagonals and all possible "broken diagonals" also give the constant sum. See figure 6.2.

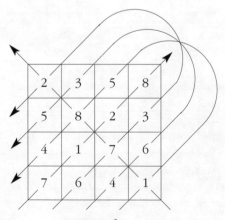

FIGURE 6.2

It is interesting that Varahamihira calls this square *kacchaputa*, 'the shell of a turtle'—does this refer to the mystical origin of the *luoshu* on the shell of a tortoise? Another early work on magic squares bears a similar title, *Kaksaputa*. This work, which also considers a fourth-order magic square (magic constant 100), is attributed to the alchemist Nagarjuna.[12] Cammann speculates that the Indians acquired their initial knowledge of magic squares either directly or indirectly from China via merchants, Buddhist missionaries, or by way of Tibetan or Muslim intermediaries; however, decisive evidence for a conclusion on this issue is lacking.[13]

The first mention of a magic square of order three in India is also clouded in legend. It is attributed to the divination efforts of a mysterious Garga who, in his writings *Gargasamhita*, recommends the use of specific third-order magic squares to pacify the nine planets.[14] See figure 6.3.

6	1	8
7	5	3
2	9	4

Sun

7	2	9
8	6	4
3	10	5

Moon

8	3	10
9	7	5
4	11	6

Mars

9	4	11
10	8	6
5	12	7

Mercury

10	5	12
11	9	7
6	3	8

Jupiter

11	6	13
12	10	8
7	14	9

Venus

12	7	14
13	11	9
8	15	10

Saturn

13	8	15
14	12	10
9	16	11

Rahu

14	9	16
15	13	11
10	17	12

Ketu

FIGURE 6.3

In this scheme, the magic square representing the Sun is the *luoshu* rotated 180 degrees. The oldest version of *Gargasamhita* dates from approximately 100 C.E.; however, its consideration of planets, *navagraha*, has been judged to have been written no earlier than 400 C.E.[15] The first documented occurrence of a third-order magic square is found in the medical thesis *Saddhayoga* (ca. 900 C.E.) where its author, Vrnda, prescribed a magic square of order three to ease the difficulties of child birth.[16] It is to be employed as a *yantra*, a mystical diagram that attracts certain powers, a talisman, while a specific *mantra* is recited. Vrnda's magic square is shown in figure 6.4 and could have been derived from the *luoshu*.

16	6	8
2	10	18
12	14	4

FIGURE 6.4

Perhaps the most famous of the medieval Indian magic squares are those of order four found engraved within Jaina temple complexes at Dudhai, Jhansi District, and Khajuraho. Paleographical dating suggests that these engravings date from the twelfth or thirteenth centuries. The Khajuraho square has a magic constant of 34. A. H. Frost, a British missionary stationed in the nearby town of Nasik, made a study of the Jaina square and published his findings for a British audience.[17] Reverend Frost termed the square "Nasik" and "pandiagonal," a term he coined to describe its particular property whereby all diagonals and broken diagonals add up to the same sum. Frost's Nasik square is shown in figure 6.5.

7	12	1	14
2	13	8	11
16	3	10	5
9	6	15	4

FIGURE 6.5

It appears that the Jains held a special reverence for magic squares. They incorporated magic squares in their religious practices and mentioned them in their hymns.[18]

The first Indian work to treat magic squares as mathematical entities was *Ganitasara* (ca. 1315 C.E.) written by the Jaina scholar, Thakkukra Pheru. He limited his considerations to the square of order four. A complete and comprehensive study of magic squares appeared in 1356 C.E. in a work entitled *Ganitakaumudi*. It was written by Narayana who devoted one chapter, *bhadra-ganita* [Mathematics of Magic Squares], to the subject.[19]

While much of the Indian work on magic squares seemed to center on squares of order four, squares as large as order fourteen and other magical configurations, such as "lotuses" and circles, were also devised. In India the square of order three has served as a talisman for good fortune that was thought to ease childbirth, help win a lover, prevent a heart attack, help establish cordial relations with a superior, insure the birth of a male heir, cure a dog bite, deflect the negative influence of planets, among other things. The square itself usually served as a *yantra* in a multifaceted ritual: one had to write the number configuration on a designated type of material, often using a special substance; one had to perform a *mantra*, special sound, or formalistic incantation; sometimes one had to sprinkle water and execute prescribed physical movements. For example, the magic square for freeing a man from prison is shown in figure 6.6.

8	1	6
3	5	7
4	9	2

FIGURE 6.6

It is an inverted *luoshu* configuration. To be effective, it was to be written on the ground 1008 times with a pen made from a kind of wood called "Aagar."[20] For ease of childbirth, Vrnda's magic square was to be gazed upon by the prospective mother while she was submerged in a bath of consecrated water. Even today, astrological counseling in India might result in a petitioner constructing one of Garga's magic squares using the wing of a peacock as a pen, writing on Bhooj Patra with Ashat Gand as ink—the Ashat having been mixed with Ganges River water.[21]

Many Indian magic squares bear Islamic numerals, which may speak to their Islamic origins or to the influence of the mogul princes who supported the use of these talisman. For example, a square that is still used to thwart the influence of a ghost is shown in figure 6.7.

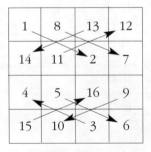

FIGURE 6.7

Outside of their occult ritual uses, magic squares also seem to have held a broader transcendental appeal to Indians. In the *Ganitakaumudi*, Narayana refers to a particular magic square of order four on several occasions. A derivation of this same square was discovered in the nineteenth century at a ruined temple complex in Gwaliar and was dated to 1483 C.E.[22] Cammann determined that the popularity of these squares lay in a pattern revealed by their numbers.[23] If the numbers of the two squares are taken in pairs following the natural progression: 1-2, 3-4, ..., a mesh of interlocking line segments results, as shown in figure 6.8.

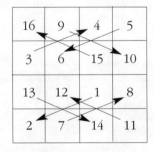

Narayana's square   Gwalior square

FIGURE 6.8

These lines move parallel to each other and alternate direction as would the threads in a weaving. Hindus conceived of the universe as a vast fabric woven by their gods. The weaving of numbers in these magic squares makes the squares themselves microcosms of the universe.

Hindus and Buddhists reinforced their belief in existence as a continuous cycle of birth, death, and reincarnation by constructing odd-order magic squares using a continuous process. It goes as follows:

1. Construct a magic square grid (for illustrative purposes, we will use a 3 × 3 grid). Call this grid the objective grid.
2. Surround the objective grid on all sides by other grids (auxiliary grids) so that a mosaic of grids are formed.
3. Place a 1 in the upper center cell of the objective grid.
4. Move up one cell and one cell to the right and place a 2 in this cell. Now since this entry is outside of the objective grid, it will be projected back to occupy the same relative position in the objective grid. (In the diagram below, circled entries are outside the objective grid.)
5. Continue through the natural number sequence in this fashion, moving one cell up and one cell to the right for each additional entry. If the entry falls in the objective square that is its final position, if not, it must be projected into the objective square.
6. When a position arrived at is already occupied by a number, the new entry will assume the position directly below its predecessor.
7. Continue the process until the objective grid is filled. One additional step will return us to the first entry, 1, closing the cycle.

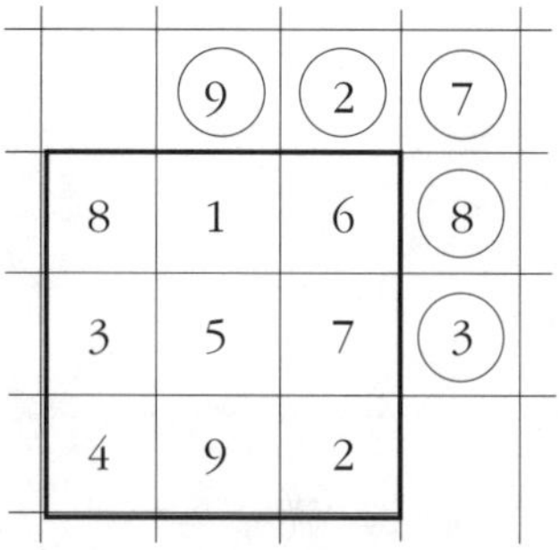

FIGURE 6.9

If one considers the natural world analogous to a magic square arrived at by this continuous process, then one achieves a place in the natural world by leaving it and then returning, a cycle of reincarnation.

Tibet

The Tibetan people living in the shadows of two major civilizations adopted and adapted many of the theories of their neighbors. In particular, Tibetan astronomy/astrology is built upon four traditions: *skar-rtsis*, 'star calculation', comes from the *Kalacakra* astronomy of India; *dbyans-char* from Indian *svarodaya*, 'divination practices'; *nag-rtsis*, 'black calculation', is based on Chinese metaphysical theories; and finally, *rgya-rtsis*, 'Chinese calculation', concerns the Chinese Shixian calendar.[24] *Nag-rtsis* was introduced into Tibet from China in the seventh century. It focused on astrological theories and practices involving *yinyang, wuxing,* and *bagua* and the "nine palaces." With these theories came the use of the *luoshu* as an astrological and divination reference. Little information is available about how Tibetan astrologers used the *luoshu*; however, most likely it was used in fortune-telling and as an occult charm. Traditional Tibetan artworks frequently contain graphic almanacs in which the *luoshu* is prominently featured.[25] See figure 6.10.

FIGURE 6.10

Japan

Early Japan borrowed cultural forms from its neighbor, China. Japanese musical theories, early art motifs, personal dress, architectural design, and city-planning principles as well as the Japanese system of written characters were all adopted from the Chinese. Daoist and Buddhist missionaries and commercial adventurers visited the islands of Japan seeking spiritual or personal gain and at the same time brought their theories and doctrines to Japan. Buddhism from China was established as a religion in Japan by the sixth century. In the seventh century, Korean Daoist priests brought astrological and calendrical texts with them into Japan. In 604, the Japanese officially adopted China's Song calendar. The metaphysical concepts of *yinyang* and *wuxing* became prevalent in Japanese cosmological thinking; however, no clear evidence indicates that the *luoshu* was an early import of note. The first official mention of this magic square in Japanese life was in the Heian period (709–1192) when, in the year 970, Tamenori Minamoto published *Kuchi-zusami*, a text for the education of young noblemen. In the twelfth chapter of this book, Minamoto presents the *luoshu* accompanied by Zhen Luan's anthropomorphic description associating the square's numbers with body parts. Although the author does not elaborate on the *luoshu's* significance or use at this time, later seventeenth-century mathematical works refer to the early use of the magic square in fortune-telling and divination practices.[26]

The study of magic squares as mathematical entities attracted Japanese scholars' interest during the Kan-ei period (1624–1643) when Chen Dawei's *Suanfa tongzong* [Systematic Treatise on Arithmetic] (1592) reached Japan. While this book was mainly a practical mathematics text stressing abacus calculation and reviewing the methods of the standard Chinese reference the *Nine Chapters on the Mathematical Art*, it also offered some mathematical challenges and puzzles including information on various types of magic squares. Japanese mathematicians seized on this information and began producing their own theories and books on magic squares. Most noticeable in this collection of works is *Kigu hosu* (1697) in which its author, Yueki Ando (1624–1704), gives methods for constructing squares of all orders three to thirty.[27] For the three-by-three *luoshu* square, Ando began with the natural square of order three (see figure 6.11a), he then exchanged the entries: 1 with 6; 3 with 8; 9 with 4, and 2 with 7. This procedure gave him a square with even-numbered entries in the corners (figure 6.11b). He then rotated the set of corner numbers 90 degrees clockwise to arrive at the *luoshu* configuration (figure 6.11c):

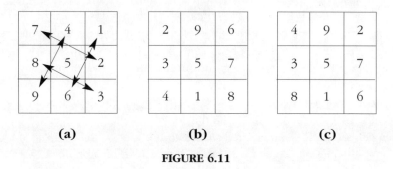

FIGURE 6.11

A later technique devised by Katahiro Takebe (1664–1739) uses an interesting transformation approach to devise the magic square of order three.[28] See figure 6.12. Takebe visualizes the natural square of order three as comprised of two cruciform subsets of numbers: one formed by the two diagonals, the other by the vertical and horizontal elements—the pivotal element for both remains the number 5. Takebe first rotates the diagonal set 45 degrees counterclockwise (figure 6.12b), he then rotates the remaining set of elements 180 degrees to arrive at the *luoshu* configuration (figure 6.12c).

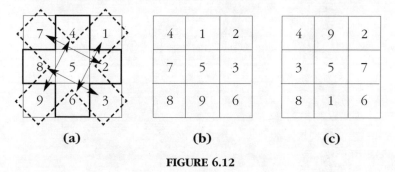

FIGURE 6.12

Both Ando's and Takebe's methods of constructing a magic square of order three stress the significance of placing even numbers at the corners of the square. In fact, if this is not done, the construction of the third-order magic square is impossible. Further, their techniques employ discrete mathematical steps focused on parts of the number configuration. Such discrete thinking seems to deviate from the continuous methods employed by other early manipulators—specifically, Indian and

Islamic thinkers—of the magic square of order three and is uniquely Japanese.

Thus it appears that by the time the *luoshu* was introduced into Japan by (most probably) Daoist emissaries, it had already lost much of its cosmological significance, and it remained primarily a device for occult purposes. A seventeenth-century image of a Japanese *luoshu* diagram associates it with palm-reading and reveals an interesting innovation in the depiction of the central entry.[29] See figure 6.13. In this illustration, the collection of circular images representing 5 is presented as a square formed by four circular images and a central image all connected by line segments. The four outer circles point to the four major compass directions—this design seems to stress the importance of centrality.

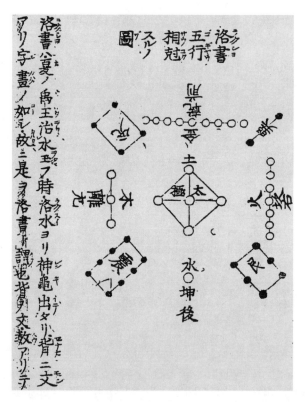

FIGURE 6.13

Later Japanese innovators of magic squares pursued purely mathematical investigations of the number configurations.

The Islamic World

Many Semitic peoples, for example, Chaldeans, Persians, and Sabaeans, had trading contacts with the Middle Kingdom long before the rise of Islam in 622 C.E.; however, little evidence exists to document their exchange of mathematical and scientific information with the Chinese or the state of their indigenous knowledge. After the rise of Islam and the establishment of stronger literary traditions, the picture becomes clearer. Islamic scholars were great synthesizers, bringing together theories and ideas from various cultures and producing new and often novel approaches to existing concepts as well as developing new ones. Islamic astrologers and mathematicians often blended Pythagorean mysticism with number knowledge and concepts obtained from the orient.

The first recorded Islamic involvement with magic squares is attributed to Jabir ibn Hayyan, the Father of Islamic alchemy.[30] Known in the West as Geber, Jabir is an enigmatic figure who is believed to have flourished within the period 875–975 and is credited with authoring over three thousand publications. Most modern scholars believe that the great majority of these writings, however, were actually written by Jabir's later followers and attributed, in true Pythagorean fashion, to their master. In the Jabiran *Kitab al-Manwazin al-Saghir* [The Small Book of Balances], a collection of one hundred forty-four treatises, one finds reference to the magic square of order three, the *luoshu*. Many Muslim scholars adopted the Greek concept of the world of material creation based on four primary elements: Water, Fire, Earth, and Air. Jabir, deeply affected by Pythagorean mysticism, uses this square to deduce the ratios in which the elements combine to form all substances. He attributes the magic square of order three and much of his proclaimed theory to Apollonius of Tyana (who lived during the first century). This Cappadocian Pythagorean philosopher, mathematician, and creator of talismans had a large following. Among the works attributed to him was *Sirr al-Khaliqa* [The Secrets of Creation], which heavily influenced Jabirian theories. Jabir associated the number 17 with harmony.[31] The choice of 17 as a number of accord seems particularly strange; mathematically speaking, 17 is undistinguished and historically, prior to this period, it bore little significance—one possible connection is that the philosopher Posidonius (135–151 B.C.E.) taught that the human soul possessed seventeen faculties. Perhaps by projecting this characteristic onto a "World Soul," Jabir arrives at this number, which is also the sum 1 + 3 + 5 + 8, numbers contained in the lower left subsquare of the *luoshu* that represent the Four Elements: 1-fire; 3-earth; 5-water, and 8-air. Further, of the remaining numbers in the magic square, when those bounded by a gnomon are summed, 4 + 9 + 2 + 7 + 6, the result is 28.[32] Twenty-eight is the second perfect number in Pythagorean tradition, and it is the number of seven planets, 1 + 2 + 3 + 4 + 5 + 6 + 7. It also enumerates the "mansions of the moon," twenty-eight regions of the heavens marking the moon's monthly path that were used as reference points by ancient astronomers and astrologers. See figure 6.14.

4	9	2
3	5	7
8	1	6

FIGURE 6.14

The mansions of the moon were also thought to be microcosmic counterparts representing the twenty-eight letters of the Arabic alphabet. These letters provided the vehicle for the Divine Word as given in the Quran and for devout Muslims were considered a form of the Divine Breath, the essence of all creation.[33] This magic square, if comprised of Arabic letters encoded to represent numbers rather then the numbers themselves, became an object for communicating sacred belief and holy passages from the Quran. In Islam, this system of *isopsephia* became known as *abjad,* a meaningless word formed from the first four letters of the Arabic alphabet. *Abjad* variants of magic squares, particularly the square of order three, became powerful talismans among Muslims and their neighbors.[34]

Jabirian theory associated the *luoshu* and its related numerology with the practices of alchemy. According to this theory, all substances possessed properties dependent on the Four Elements. These properties existed in fixed ratios centered on the number 17 and if the ratios were altered a transmutation of the substance could take place.[35] Alchemists believed that metals possessed two sets of such properties, "inner qualities" and "outer qualities." Thus an analysis in the *Book of Balances* notes that lead contains 3 parts of coldness and 8 parts of dryness as outer qualities and 1 part of heat and 5 parts of humidity as inner qualities, whereas gold possesses 3 parts of heat and 8 parts of humidity as outer qualities and 1 part of coldness and 5 parts of dryness as inner qualities. Note that the sum of the parts in each case totals 17 parts. Further, alchemists held that the proportion of these qualities could be changed and, in theory, lead could be converted into gold. This principle became the basis of the fabled alchemistic quests for wealth by turning common metal into gold.

Jabir also proclaimed the *abjad* version of the *luoshu* a valuable aid to childbirth, a belief similar to one held in India. The magic square of order three and other magic squares adopted or devised by Muslims became therapeutic devices when used to adorn plates, bowls, and cups to whose surfaces they imparted their healing powers. In the later nineteenth century, porcelain factories in China still manufactured and exported these "medicine plates."[36]

The first published set of magic squares appeared in the encyclopedia *Rasa' il* (ca. 989), a forty-eight volume series compiled by the *Ikhwan as-Safa,* a Muslim sect, popularly known as the "Brethren of Purity."[37] This sect, which was centered in Basra—a thriving seaport at this time and a marketplace of foreign influences—believed in the purifying power of knowledge. Ikhwanian writings and teachings represented a Gnostic effort to reconcile Hellenistic beliefs with the teachings of the Quran. Pythagorean and Neoplatonic mystical concepts were dominant in the Brethren's rationales. They viewed geometric shapes as personalities bearing special attributes. The triangle represented harmony. The square represented stability and promoted the Four Element theory and thus a special status for the number 4 (reminiscent of the Chinese cosmological reverence for the number five). Of this fourfold division of Nature, the Ikhwans wrote:

> God himself has made it such that the majority of the things
> of Nature are grouped in four such as the four physical natures
> which are hot, cold, dry and moist; the four elements which are
> fire, air, water and earth; the four humours which are blood,
> phlegm, yellow bile and black bile; the four seasons . . . , the
> four cardinal directions . . . , the four winds . . . , the four directions
> envisaged by their relation to the constellations; the four products
> which are the metals, plants, animals and men.[38]

For the Ikhwans, as for the Pythagoreans, number appeared to be the key for understanding the relationships upon which the physical and spiritual world functioned. They developed their own system of *abjad* for the numerical symbolism of letters and devised numerical categories to clarify relationships. In particular, the Brethren divided all beings and objects into nine states, since 9, the last digit in the decimal numeration system, closes a cycle and symbolically brings an end to the series of numbers. The categories are as follows:

1. Creator, who is one, simple, eternal, permanent.
2. Intellect, which is of two kinds: innate and acquired.
3. Soul, which has three species: vegetative, animal, and rational.
4. Matter, which is of four kinds: matter of artifacts, physical matter, universal matter, and original matter.
5. Nature, which is of five kinds: celestial nature and the four elemental natures.
6. Body, which has six directions: above, below, front, back, left, and right.
7. The sphere [of the Universe], which has its seven planets.
8. The elements which possess eight qualities: four qualities combined two by two:

    Earth—cold and dry
    Water—cold and wet
    Air—warm and wet
    Fire—warm and dry

9. Beings of the world, which are the mineral, plant, and animal kingdoms, each having three parts.[39]

The first four numbers in this series and the entities in the first four categories were, by the Pythagorean concept of *tetractys,* considered to be pure, universal beings since, within their sum, they contained all numbers. The other beings are compound objects.

Evidence in the *Rasa'il* indicates the Ikhwan Brethren had knowledge of magic squares up to and including order nine. The Arabic term for a magic square is *wafq al-a'dad,* 'the harmonious disposition of numbers', and the Brethren considered their magic squares "small models of a harmonious universe." Their magic square of order three is presented as the *luoshu* rotated 90 degrees clockwise. Its construction is described in terms of chess moves—"First, two knight's leaps, then a pawn step"—which reveals that, by this time, Islamic scholars had experimented with the structures of magic squares and had devised their own techniques for developing them. However, an examination of the *Rasa'il* magic squares reveals no dominating technique for their construction but rather an eclectic mix of methods. Perhaps the most interesting and innovative square in this series is the one of order seven, a concentric bordered magic square. Its author apparently took the middle numbers from the sequence 1 to 49, namely, 21 to 29, and arranged them into a magic square of order three which he then used as a core for further construction.[40] Next, he chose the sequences 30 to 37 and 13 to 20 and paired elements of each to form complementary pairs whose sum was 50. He distributed these as a border around the central core. Finally, he employed the remaining numbers in twelve complementary pairs to form the outer border. The resulting square was comprised of a core with two concentric borders—a magic square within a magic square within a magic square! See figure 6.15.

47	11	8	9	6	45	49
4	37	20	17	16	35	46
2	18	26	21	28	32	48
43	19	27	25	23	31	7
38	36	22	29	24	14	12
40	15	30	33	34	13	10
1	39	42	41	44	5	3

FIGURE 6.15

Subsequent Islamic scholars expanded this body of knowledge about magic squares and their uses. The versatile scientist-philosopher Thabit ibn Qurra (836–901) wrote on magic squares. Ibn Sina (980–1037), an Isma'ili, known in the west as Avicenna and respected for his medical knowledge, used magic squares. Al-Ghazali of Tus (1058–1111), primarily known as a theologian, mentions them in his work. Ibn al-Lubudi (b. 1210), a Syrian physician, astronomer, and mathematician, while considered a medical author, wrote an essay on magic squares.[41] Abu'l-Abbas al-Buni (d. 1225), a north African occultist, wrote three books on the use of magic squares as talismans. His most comprehensive but cryptic writings on the subject are found in *Shams al-ma'arif al-Kubra* in which he labels the *luoshu* "*Izra'il*," 'Angel of Death'.[42] Several anonymous Persian manuscripts from the thirteenth century attest to the widespread interest in magic squares in the Islamic world.[43] An examination of this literature reveals two features worthy of further note: Islamic scholars developed a continuous technique to generate odd-order squares, and they believed that magic squares were truly magical devices, particularly useful in the mystical "sciences" of medicine, alchemy, and astrology.

Early Islamic works indicate a facility with a diamond technique for devising odd-order squares. In this procedure the numbers are written in consecutive order in a series of right-slanting diagonal lines intersecting the respective magic square grid. The numbers in each line reflect the order of the square; thus, a third-order square will have three entries on each diagonal line. In this process all numbers falling inside the grid retain their position, those outside the boundaries of the grid are projected back within the opposite side of the grid. Examples of this process are demonstrated for third- and fifth-order magic squares in figure 6.16.

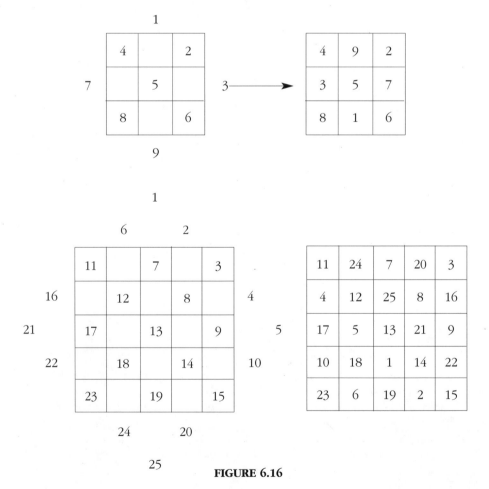

FIGURE 6.16

Once established, this method provides a simple, easily remembered algorithm for constructing odd-order squares. But by the thirteenth century, Persian texts indicate that their authors had devised another method, one that perhaps was more psychologically appealing. This method began inside the square, proceeded in a continual fashion, and ended at the point where it started, closing the cycle.

In an empty grid for an odd-order magic square, the author placed the number 1 in the cell directly below the central cell. Then he continued downward and to the right, placing a 2 in the next available cell. If the cell lies within the grid, the number retains its position; however, if it falls outside the grid's boundary then the number is projected into the farthest empty cell within the row or column marked by the number from which the process continues. When a target cell is already occupied, the enumeration breaks from its diagonal path, drops down one cell directly below the previously placed number (a "break move") and begins again. This process is followed until the square is completed. An example of this method is shown in figure 6.17 where circled entries are situated as a result of a break move.

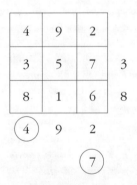

FIGURE 6.17

Conceptually, this system works as if the magic square were constructed on a doubly curved surface which behaves both like a vertical cylinder—that is, the right side of the square curves to meet the left side—and, a horizontal cylinder where the top curves to meet the bottom. Geometrically speaking, such a surface does exist; it is the donut-shaped "torus"; however, it is doubtful if the creators of this technique relied on such a geometric insight.[44]

In Islamic tradition, magic squares were interesting not for their mathematical properties but for their metaphysical potential as communicators of religious and occult messages. Scholars created many *abjad* systems for encoding words and textual passages into numerical or letter equivalents.[45] The system used by Ibn Sina placed principal importance on the first nine letters of the Arabic alphabet because it was believed these nine letters, known as the "Nine Letters of Adam," were the primary instruments through which God first communicated to humans. These nine letters established a special hierarchy of relationships among the realms of creation. All the other letters of Ibn Sina's system were formed by adding or multiplying together the numbers corresponding to these first nine levels or states of being. Here is Ibn Sina's code:

A = 1 = *al-Bari'*: Creator
B = 2 = *al-'aql*: Intellect
J = 3 = *al-nafs*: Soul
D = 4 = *al-tabi'ah*: Nature
H = 5 = *al-Bari' (Bi'l-idafah)*: Creator in relation to what is below
W = 6 = *al'aql (bi'l-idafah)*: Intellect in relation to what is below it
Z = 7 = *al-nafs (bi'l-idafah)*: Soul in relation to what is below it
H = 8 = *al-tabi'ah (bi'l-idafah)*: Nature in relation to what is below it
T = 9 = *al-hayula'*: material world having no relation to anything below it
Y = 10 = 5 × 2 = *al-ibda'*: the plan of the Creator
K = 20 = 5 × 4 = *al-takwin*: Structure transmitted to the created realm
L = 30 = 5 × 6 = *al-amr*: the Divine Commandment
M = 40 = 5 × 8 = *al-khalq*: the created Universe
N = 50 = M + Y = the twofold aspect of *wiyud* (being)
S = 60 = M + K = the double relation to *khalq* and *takwin*
'ayn = 70 = L + M = *al-tartib*: chain of being impressed upon the Universe
S = 90 = L + M + K = the triple relation to *amr, khalq* and *takwin*
Q = 100 = 2Y = S + Y = *ishtimal al-jumlah fi'l-ibda'*: The assembly of all things in the plan of the Creator[46]

The *abjad* version of the *luoshu* talisman that Jabir discusses is shown in figure 6.18.

Variations of this square have served Muslims as religious mandalas, meditative devices, and occult talismans and amulets. The square's magic sum of 15, which for the Chinese represented "human being perfected," has been culturally transformed in the Islamic context to reflect the phrase "O man" or "O Perfect Man" referring to the Prophet Muhammad. A configuration formed from the square's central row and column—a cross of the odd numbers—is considered a harbinger of bad

The Magic Square of Order Three in Other Cultures 101

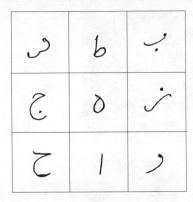

FIGURE 6.18

luck or evil, whereas a configuration of the four corner numbers—the even numbers—is thought to be a powerful talisman for good luck. The Arabic letters for the four corners, if read from right to left, form the word BEDUH, the "word of power" as handed down by Adam.[47] This word itself, or its abbreviations—just the letter B or the numerical equivalents 2, 4, 6, 8—are believed to be powerful talismans that protect travelers, babies, and postal letters and packages in transit. In some Islamic countries today, one finds packages with "2," "4," "6," "8" written in their corners or postal letters bearing an extraneous "B" written under the address as added postal insurance. The BEDUH tradition is believed to predate the Quran. Among its other attributes, the Islamic *luoshu* is credited with insuring that one finds love; helping one secure a mate; preventing childhood fears; curing headaches, stomach ailments, fevers, and epilepsy; preventing theft, attacks by bandits, and poisonings; and helping one find lost objects. These different applications may have required writing the square on a specific surface: a part of the human body, for example, the forehead, palms, or fingernails; using a specific substance, such as blood, to draw the square, and going through a physical ritual or performing an accompanying chant.

The magic square of order three was highly revered throughout the lands of Islamic influence. In medieval times it was inscribed on small coin-like amulets and widely distributed. A large collection of these amulets can be viewed at the Indonesian National Museum in Jakarta where the *luoshu* can also be found adorning antique prayer rugs and sarongs. Travelers to Palembang on the southern coast of Sumatra in the late nineteenth century still found these amulets being sold, engraved on the face of a ring. At about the same time, young children in the

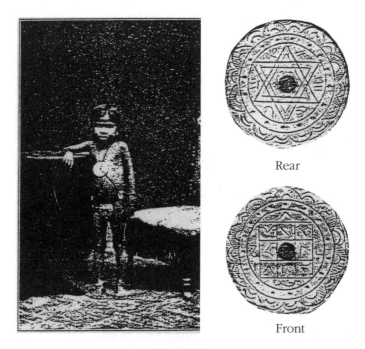

FIGURE 6.19

Celebes were wearing large silver amulets suspended from their necks and engraved with the magic square.[48] See figure 6.19.

Walter Skeat, writing on the magic practices of the Malay peninsula (now West Malaysia and Singapore), provides some details on just how these amulets were constructed and used.[49] In many cases, when a person sought a specific favor through the use of an amulet, that amulet was personalized with an encoded inscription of the person's name. Each letter of the name would be assigned a numerical value according to the system used. To seduce a woman, the names of the woman and her would-be seducer were required; if a married couple desired a child both their names were necessary. By encoding the letters of the names, the "key number" was arrived at, at which point it had 12 subtracted from it and then the remainder was divided by 4; the result was placed in an empty cell in the waiting magic square array. Usually this cell is the central one in the *luoshu* configuration. Skeat describes the use of a magic square talisman intended to secure wealth for its patron.[50] This talisman is shown in figure 6.20.

The patron's name (in numbers) is written in the empty cells and his or her desires are listed in the space between the squares. The charm is

The Magic Square of Order Three in Other Cultures 103

FIGURE 6.20

then carried on the patron's person. Its preparer must undertake ritual ablutions as though he were attending prayers at a mosque. (The preparer is specified as being male.) He must also clean and purify the surface on which the charm is being written. He further prepares himself by praying, reading the *al-Fatihah*, the first chapter of the Quran. Then he recites fifteen repetitions of the "Throne verse," verse 255 of chapter 2, and twenty-five repetitions of chapter 112. After this extensive preparation, he writes the charm. Some magic square charms were inexpensively mass-produced as block prints on green paper, a color often associated with Islam.

While generally referred to as magic square talismans or charms, many of these number configurations were not magic squares in a true mathematical sense. In the charm for wealth described above, the border rows and columns of the left square each add up to the sum of 14 as do the central number pairs from opposite sides; for the right square, the pattern is the same but the constant sum is 19. Obviously the numbers 14 and 19 had significance within this mystical context. The use of such "border squares" in securing desires and in fortune-telling was widespread. Westermarck reported the use of such squares in Fez, Morocco in the 1920s.[51] He observed squares used for fortune-telling, with the patron's number inserted into the center cell. One such square is shown in figure 6.21.

4	9	1
8		6
2	5	7

FIGURE 6.21

The *Ilm al-asrar*, 'the science of secrets', was well-known and practiced by the Fulani people of north Africa. Muhammad ibn Muhammad, an eighteenth-century mathematician, astrologer, and mystic who lived in Katsina (northern Nigeria) published a book on magic squares in 1732. In his writing, Muhammad counsels his students of *abjad*, advising:

> work in secret and privacy. The letters are in God's safekeeping. God's power is in his names and his secrets, and if you enter his treasury you are in God's privacy, and you should not spread God's secrets indiscriminately.[52]

Noted for their acuity as merchants and breeders and traders of cattle, the Fulani were also ardent believers and users of magic square charms. They even incorporated magic squares into games in which the challenge was to complete a magic square by placing stones or pieces of dung in the empty cells of a grid drawn on the ground—the magic sum would be specified. Two squares in popular use among the Fulani and Islamic peoples in general are the magic square of order three whose sum gives the number of Allah and the square of order four whose magic constant is the number of the "Four Angels."[53] The numerical equivalents of these squares are shown in figure 6.22.

21	26	19
20	22	24
25	18	23

Allah (66)

23134	23137	23143	23127
23142	23128	23133	23138
23129	23145	23135	23132
23136	23131	23130	23144

Four Angels (92541)

FIGURE 6.22

One of the most dramatic descriptions of the *luoshu* magic square in the Islamic world comes from the early nineteenth-century traveler Edward Lane.[54] Lane, while visiting Cairo, saw a street magician put a youth into a trance.[55] The magician drew a version of the *luoshu* on the young man's right palm. See figure 6.23. He then placed a large drop of black ink in its center. Required to gaze into the drop, the youth fell into

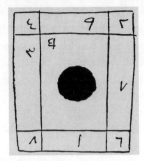

FIGURE 6.23

a deep trance and, by the power of his master's suggestion, saw scenes of buildings and people within this drop of ink. He vividly described his visions to the amusement of the street audience.

Stylized calligraphic renderings of magic squares are visually, psychologically, and esthetically appealing, so for a devout Muslim beholder, their spiritual impact is greatly enhanced. A north African talisman for wealth and good fortune is shown in figure 6.24(a).[56] The numbers enclosed in the artistic design correspond to the "Allah square" of figure 6.22. Islamic mystical charms based on magic squares also represented many religious figures. The square illustrated in figure 6.24(b) pays homage to Allah, the Prophet Muhammad, and the Prophet's four successors, the caliphs: Abu Bakr, Umar, Uthman, and Ali.[57]

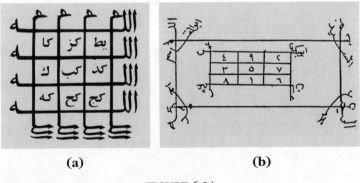

(a) (b)

FIGURE 6.24

While magic squares had many uses within the Islamic tradition, astrology eventually became a major focus. Magic squares were associated with the planets and served as the basis for planetary amulets. Much of Islamic astrological theory was adopted from Harranian beliefs.

Harran was a Syrian town on the upper Euphrates that served as a repository of ancient Babylonian traditions well into the tenth century of the Christian Era. Harranian theories associated the planets with specific colors, metals, geometric shapes, and numbers and had temples and idols dedicated to each planet. Table 6.1 lists these relationships: [58]

TABLE 6.1

Planet to which the Temple was dedicated	Metal of which the God's image was made	Associated Colour	Geometrical Structure of Temple	Number of Steps to the throne of each Idol
SATURN	LEAD	Black	Hexagonal	9
JUPITER	TIN	Green	Triangular base: roof and angles pointed	8
MARS	IRON	Red	Oblong	7
SUN	GOLD (image hung with PEARLS)	Yellow	Square	6
VENUS	COPPER	Blue	Triangular (with one side longer than the other two)	5
MERCURY	An alloy of all the metals, including Kharsini (Chinese Iron). The hollow interior was filled with MERCURY - thus imparting to the Image of the Spirit f the Planetary deity	Brown (At the Wednesday service a Brown youth who was a good scribe was slain, quartered, the quarters separately burnt, and the ashes thrown in the face of the image	Hexagonal, with a Square interior	4 (circular)
MOON	SILVER	White	Pentagonal	3

Muslims rejected the idolatrous aspects of the planetary theory but did adopt other correlations, for example, a planet's connection to a specific metal. The Persian astrologer Abu Ma'shar (d. 886) mentions the planets and their respective metals in his work. One of Islamic Spain's greatest scientists was al-Majriti, Abu-al-Qasim Maslamah ibn Ahmad al-Faradi (d. 1007). Born in Madrid, al-Majriti, who may have served as a court astrologer, wrote works on spherical geometry, astrology, and commercial arithmetic.[59] He had a large following of disciples, one of whom, Ibn al-Samh (d. 1035), wrote a treatise, *Book of the Plates of the Seven Planets,* now lost in its original Arabic but reproduced in the Spanish work, *Libros del Saber de Astrologia.* Ibn al-Samh also associated each planet with a certain metal. Alfonso X, the King of Castile, ordered the publication of a compendium of astrological and cosmological beliefs. The work, first written in Spanish and then translated into Latin, was based mainly on the text, *Ghajat al-Hakim* [The Aim of the Sage], attributed to al-Majriti. It appeared in 1256 under the Latin title *Picatrix* and was widely distributed throughout Europe. The *Picatrix* introduced Latin Europe to the mystical practices of Islamic astrology including the use of planetary amulets involving magic squares; however, it lacked detailed explanations.[60] About thirty years before, al-Buni in North Africa had published a complete description of such amulets in which Saturn was represented by a third-order magic square, the *luoshu*; Jupiter, a fourth-order square; Mars, a fifth-order square; the Sun, a sixth-order square; Venus, a seventh-order square; Mercury, an eighth-order square; and the Moon, a ninth-order magic square. These amulets were each to be made of a different metal as per the Harranian tradition. Lead, the "father of all metals," was to be the substance upon which the *luoshu* was inscribed. By the fifteenth century, this knowledge found its way into Latin Europe. A manuscript of this period, now in the Jagiellonian Library in Krakow, contains the earliest known set of magic squares in the Latin world.[61] They are the seven planetary magic squares of the Islamic tradition.

Magic Squares in Latin Europe

While the appearance of the *Picatrix* was a major influence on stimulating a broader European interest in magic squares, other information on the number configurations and their uses was also finding its way into Latin Europe.[62] Abraham ben Meir ibn Ezra (1090–1167), alternately known as Abenezra, was a Spanish-Jewish philosopher, biblical commentator, astrologer, mathematician, and translator of Arabic works into Hebrew. He resided and taught in Toledo but traveled widely, journeying to North Africa, Italy, France, and even England.[63] Abraham was a Neoplatonist who believed strongly in astrology and has been credited with writing over fifty works on the subject.[64] His works included a magic square of order three with its entries comprised of characters of the Hebrew alphabet used to represent numbers. See figure 6.25.

FIGURE 6.25

Upon numerical interpretation, it turns out that this square is the *luoshu*. This magic square was readily adopted into the existing Jewish Kabbalistic tradition as a mystical symbol and found its way into Jewish communities across Europe. The historical introduction of magic squares into Europe has often been attributed to Manual Moschopoulos (ca. 1300), a Greek grammarian who lived and taught in Constantinople. Moschopoulos [the "little calf"] did write a treatise on magic squares but it appears that it was little known during his time and only drew mathematical attention when it was rediscovered in the seventeenth century by Philippe de la Hire who found it in the Royal Library in Paris.[65] In his studies of fourteenth-century European mathematics, mathematical historian Kurt Vogel notes the existence of a puzzle-type problem based on the structure of the *luoshu*.[66] However, David Singmaster, in his research of mathematical puzzles, has traced the problem back to the twelfth century.[67] It can be found in *Annales Stadenses* (ca. 1240) where the version goes as follows:

There were three brothers at Cologne, who had nine casks of wine.
The first cask contained 1 bucket, the second 2, the third 3, the fourth

4, the fifth 5, the sixth 6, the seventh 7, the eighth 8, the ninth 9.
Divide this wine equally among these three without breaking any casks.

and the answer is provided:

To the oldest, I give the first [cask], fifth and ninth, and he has 15 buckets. To the middle one, I give the third, fourth and eighth, and he likewise has 15. So to the youngest I gave the second, sixth and seventh; and thus he also has 15, the wine is divided and the casks are not broken.[68]

Singmaster notes that this is a variation of wine division problems as found in Alcuin of York's *Propositiones ad acuendos juvenes* (ca. 775).[69]

By at least as early as the twelfth century, the magic square of order three was known and appreciated, in some sense, in Christian Europe. However, it was the luoshu's supposed occult or magic power (as embodied in theories of planetary influences or amulet magic) as opposed to its mathematical properties that made it widely popular.

Among the early Renaissance writers on applied arithmetic was the Florentine reckoning master, Paolo Dagomari dell'Abaco (ca. 1281–1374). Dagomari was also known as an astronomer and astrologer. In his *Trattato d'Abbaco* [Treatise on Computing], he published the magic squares for the sun and the moon.[70] About 150 years later, Luca Paciolo (ca. 1445–1509), noted mathematics teacher and compiler, put together a collection of eighty-one mathematical problems. The collection was compiled during the years 1496–1508 and entitled *De viribus quantitatis*. Problem 72 in this series contains the seven planetary magic squares. Paciolo considers the squares mathematical curiosities; he merely mentions their astrological and magical significance without exploring the subjects further. Knowledge of magic squares and their association with the power of the planets was reaching a wider and wider European audience. In 1514, the German artist Albrecht Dürer (1471–1528) published an engraving, *Melancholia I*. The print depicted the characteristics of melancholy, one of the temperaments associated with the theory of the "four humors"—an excess of black bile, which resulted in melancholy.[71] See figure 6.26. Dürer's scene contains many symbols which reflect on the presence and influence of melancholy. One of these symbols is a magic square of order four, the square of Jupiter with its numbers rearranged so that the date 1514 is made prominent in a lower central position. In astrology, the planet Saturn conveys

FIGURE 6.26

melancholy, and the planet Jupiter counteracts Saturn's influence; therefore, the Jupiter magic square is intended to lessen the effects of melancholy.[72] Dürer possessed a keen interest in mathematics and during the years 1505 to 1507 he traveled widely in Italy studying the new techniques of geometric perspective. In 1506 he visited Bologna where he met with the mathematicians Scipione de Ferro (1465–1526) and Luca Paciolo. Perhaps it was from this contact that he learned of magic squares.[73] Dürer's square of order four was the same as the one contained in the fifteenth-century Krakow manuscript.

But magic squares would make their greatest intellectual impact on the European scene in the realm of the mystical and the occult. Magic squares were readily appreciated for their intriguing mathematical properties by Jewish mystics of the Kabbala. The planetary squares were transcribed into Hebrew characters and revered for their magical power as talismans. See table 6.2 for the conversion values of Hebrew characters.[74] One of the tasks the Jewish mystics set for themselves was to seek out the names of God and, in particular, to find the *Shem Hameforash*, 'Ineffable Name', which would insure its knower supreme power. The one name acknowledged by all Jews for their God was YHWH [Jehovah] but it was so sacred, that for the devoutly religious Jew, it was unspeakable.[75] The *Tetragrammation*, YHWH, was given in circumlocutional forms one of which was the *luoshu* magic square since YH was the shortened form of YHWH and it possessed the numerical value 15. Further the value of the expanded Tetragrammation: YWD; HA;

TABLE 6.2

Figure.	Names.	Corresponding Letters.	Numerical Power.
1	Aleph	- - -	1
2	Baith	B	2
	Vaith	V	- -
3	Gimmel	G	3
4	Daleth	D	4
5	Hay	H	5
6	Wav	W	6
7	Zayin	Z	7
8	Cheth	Ch	8
9	Teth	T	9
10	Yood	Y	10
11	Caph	C	20
	Chaph	Ch	- -
12	Lamed	L	30
13	Mem	M	40
14	Noon	N	50
15	Samech	S	60
16	Ayin	- - -	70
17	Pay	P	80
	Phay	Ph	- -
18	Tzadè	Tz	90
19	Koof	K	100
20	Raish	R	200
21	Sheen	Sh	300
	Seen	S	- -
22	Tav	T	400
	Thav	Th	- -

4	9	2
3	5	7
8	1	6

=

ד	ט	ב
ג	ה	ז
ח	א	ו

FIGURE 6.27

WAW; HA, designated a numerical value of 45, the total of the *luoshu* entries. See figure 6.27. Kabbalists also associated the divine name with the four elements of medieval European cosmological belief: where Y, *yod* = Fire; H, *he* = Water; W, *vau* = Air, and the second H, *he* = Earth. See figure 6.28 for the magic squares of the four elements. The square for Fire is the *luoshu*. Squares for the remaining three elements are permutations of the *luoshu*.

Among the Kabbalist's number configurations, the one that is most reminiscent, in a functional sense, to the Chinese cosmological interpretation of the *luoshu* is the *Aiq Beker*, the 'Kabbala of the Nine Chambers'. This is a nine-cell square configuration where each cell contains three Hebrew letters. The Kabbala of the Nine Chambers is not a magic square in the mathematical sense but was considered quite magical. This square is comprised of the twenty-two characters of the Hebrew alphabet and five alternate or final forms for five of the characters.[76] These letters or characters are divided into three categories each containing nine letters. The first nine, associated with the numbers 1 to 9, represent the divisions of the world as ruled by nine orders of angels. The second group of nine, numbers 10 to 90, represent the things that are in the nine circles of heaven. The final set of nine consisted of the last four letters of the Hebrew alphabet and the final forms for K, M, N, P, and Tz. Each represents, respectively: the numbers 100 to 400 and the four elements: earth, air, fire, and water; the numbers 500 to 900 and the five unions of bodies of the Kabbalah tradition.[77] Figure 6.29 presents a translated Kabbala of the Nine Chambers.

Within the configuration, the letters of each chamber or cell were linked in an occult manner and could be interchanged as could their numerical values. This Kabbalic method allowing permutations was called *Temurah*. Each chamber held a special mathematical and occult

The Magic Square of Order Three in Other Cultures 113

ו	א	ח
ז	ה	ג
ב	ט	ד

6	1	8
7	5	3
2	9	4

Water

ד	ט	ב
ג	ה	ז
ח	א	ו

4	9	2
3	5	7
8	1	6

Fire

ו	ז	ב
א	ה	ט
ח	ג	ד

6	7	2
1	5	9
8	3	4

Earth

ב	ז	ו
ט	ה	א
ד	ג	ח

2	7	6
9	5	1
4	3	8

Air

FIGURE 6.28

significance and the chambers themselves gave rise to a symbolic code of nine characters, one associated with each chamber. These characters, shown in figure 6.30, served as vehicles for the conveying of mystic messages.

FIGURE 6.29

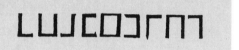

FIGURE 6.30

While the practices of Kabbala were known in Europe from very early times, the expulsion of Spain's Jews in 1492 resulted in a great influx of Kabbalic lore into the Latin kingdoms. Christian mystics, metaphysicians, and astrologers were widely attracted to this new mysterious knowledge. Johannes Reuchlin (1455–1522), a German humanist known by the name of Capnion, became the first non-Jew to publish a work on the Kabbala. His *De arte cabalistica* appeared in 1517. The work considered Kabbalic theories and numerology. Reuchlin's concern with the mystical properties of numbers led to his being described as "Pythagoras reborn." He knew of and worked with magic squares.

The man who was to formalize the theory of planetary magic squares within the occult community of Europe was Henricus Cornelius Agrippa von Nettesheim (1486–1536). Agrippa was a German scholar and mystic, a practitioner of "white magic" who would eventually serve as the model for Christopher Marlowe's Doctor Faustus. In 1531, Agrippa published *De occulta philosophia*, a manual of occult beliefs and practices.[78] The twelfth chapter of the second book of this work is entitled "Of the tables of the planets, their virtues, forms, and what divine names, intelligences, and spirits are set over them" and contains the seven planetary magic squares (tables) in both numerical and

The Magic Square of Order Three in Other Cultures

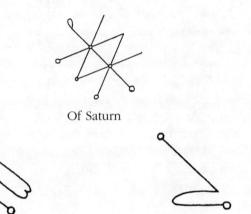

4	9	2
3	5	7
8	1	6

The table of Saturn in his Compass

In Hebrew notes

The Seals of Characters

Of Saturn

Of the Intelligence of Saturn

Of the Spirit of Saturn

FIGURE 6.31

Kabbalistic form together with various symbols associated with the planets. The *luoshu* appears as the magic square for Saturn; its seal can be recognized as the *yubu*.[79] See figure 6.32.

Accompanying text describes its talismanic powers:

> They say that this table [magic square] being with a fortunate Saturn engraven on a plate of lead, doth help to bring forth, or birth, and to make a man safe, and powerful, and to cause success of petitions with princes, and powers: but if it be done with an unfortunate Saturn, that it hinders buildings, plantings, and the like, and casts a man from honors, and dignities, and causes discords, and quarrelings, and disperses an army.[80]

A preliminary version of *Occulta* completed in 1511 did not include material on magic squares. Agrippa resided in Italy during from 1511 to 1518, and it was probably there that he came into contact with Kabbalistic circles and acquired information on the planetary squares that were then added to the later edition of his work. His work became one of the most widely distributed books of the sixteenth century and since its appearance, *De occulta philosophia* by Corneluis Agrippa has remained the paramount reference on occult beliefs associating the *luoshu* with the planet Saturn.

In 1539, Girolamo Cardano (1501–1576), Italian mathematician and acknowledged astrologer, published *Practica arithmetice et mensurandi singularis* in which he gave the seven planetary magic squares as shown in figure 6.32. Cardano, however, reversed the order of the squares, which suggests that he arrived at the information independently of Agrippa.[81] Finally in 1567, a third work associating magic squares with the planets appeared. It was entitled *Archidoxa Magica* and is attributed to Paracelsus [Theophrastus Bombastus von Hohenhiem] (1493–1541) who, during his lifetime, was widely recognized as a physician, astrologer, theologian, and mystical thinker. The seventh book of this treatise describes magic square talisman for each of the planets. Paracelsus's ordering of the seven planets agreed with that given by Agrippa. This book also attracted a large following.

Under the influence of these writings, many practitioners of astrology were now drawn into the use of magic squares and magic square talismans. Mystical experimenters such as Francesco Giorgi (1466–1540) and the British Neoplatonist John Dee (1527–1608), scientific advisor to the Elizabethan court, and Dee's colleague Robert Fludd (1574–1637) actively pursued this new numerological dimension in astrology

The occult influence of the *luoshu* was most obvious in its guise as a talisman, inscribed on lead and worn on one's person to deter the effects of the plague.

A Mathematical Interest in Magic Squares

By the dawn of the sixteenth century, the occult attraction of magic squares was strong in Europe, and their appeal as mathematical entities was beginning to grow as well. Mathematicians began to consider them as problems in mathematical arrangement and examined their workings. The German Rechenmeister Adam Riese (ca. 1489–1559) demonstrated the use of the Persian continuous method to construct odd-order magic

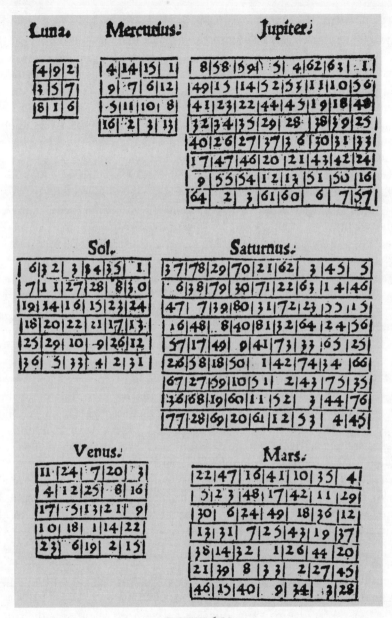

FIGURE 6.32

squares in the 1522 edition of his *Rechnung auff der Linien und Federn*. Also in Germany Michael Stifel (1486–1567) experimented with bordered squares, publishing some of his findings in his *Arithmetica integra* (1544). Such squares are still known in Germany today as Stifel squares [*Stifelsche Quadrate*].

Unaware of German developments with magic squares, French mathematicians were undertaking their own investigations. Claude-Gaspar Bachet de Meziriac (1581–1638), while studying the formation of Agrippa's odd-order magic squares, stumbled upon the "diamond technique" of construction originally employed by the Chinese. Bachet published his method in a collection of mathematical problems, *Problemes plaisants et delectables* (1624). The method was hailed as a new and exciting discovery and has since borne the name of Bachet de Meziriac. But many realized that this technique was slow and inefficient. In 1688, a returning diplomatic envoy to the Court of Siam, Simon de la Loubère, amused himself on shipboard by constructing magic squares. He used Bachet's diamond technique and attracted the attention of a fellow passenger, a French physician by the name of Vincent who was returning from a visit to Surat on the coast of India. At Surat, Vincent had learned the Hindu continuous method of constructing magic squares that he now taught to his delighted traveling companion. In 1691, Simon de la Loubère published an account of his travels including a chapter that introduced the Hindu technique for constructing magic squares.[82] This new method was readily adopted in Europe.

La Loubère's book, *Du Royaume de Siam,* in essence a travel adventure, became very popular, and while it informed its readers of the exotic ways of the Orient, it also introduced them to magic squares. One reader, the Abbé Poignard, Grand Canon of Brussels, became an enthusiastic student of magic squares and embarked on an extensive investigation of their properties. Abbé Poignard published his findings in *Traité des Quarrés sublimes* (1704). This text was reviewed for the Royal Academy of Sciences by Philippe de la Hire (1640–1718) and inspired him to become involved with magic squares. Hire published a method of creating new squares from two given squares. His theory appeared in the *Mémoires* of the Royal Academy in 1705. It attracted further interest in the mathematics of magic squares and resulted in Joseph Sauveur publishing "Construction générale des quarrés magiques" in the *Mémoires* of 1710. In a letter to Gottfried Wilhelm Leibniz dated 12 November 1712, Pierre Varignon, one for France's leading mathematicians and an academician, noted that he had just spent two months examining a book on magic squares for the Academy. This was probably Sauveur's work. From all this activity, the new method, an alternate to Bachet's technique for construction magic squares emerged—it was called the "Method of La Hire" and to this day the name has stuck.

Somehow the contributions of Sauveur, Poignard, La Loubère, and the original Indian inventors of the method were ignored.

Bernard Frenicle de Bessy (1605–1675), an accomplished amateur mathematician and number theorist, worked with doubly-even magic squares. A year after his death, Frenicle's book on magic squares, *Traité des Triangles Rectangles en nombres*, appeared. Frenicle calculated that there existed exactly 880 magic squares of order four and published a compilation of them.[83] In 1750 d'Ons-le-Bray contributed a memoir to the Royal Academy on the construction of higher order magic squares devised using Islamic techniques. He prefaced his findings by commenting on the computation of magic squares as a diversion for staving off boredom. The American statesman and natural scientist, Benjamin Franklin, constructed magic squares during the years 1736 and 1737 while serving as a clerk in the Pennsylvania Assembly. He found working with the squares reduced the tedium of his job. In January 1750, Franklin corresponded with a British friend, a James Logan, on the subject of magic squares.[84] Later when visiting England in 1769, Logan gave Benjamin Franklin a copy of Frenicle's book and told him about the remarkable French contributions to the subject. He also learned that English mathematicians had done little in this field. It would take the discoveries and enthusiasm of A. H. Frost to arouse British interest in the topic.[85]

By the nineteenth century, the magic square that had dominated early Chinese cosmological thinking, found its way into alchemy and astrology, and served as a centerpiece for medieval occult numerology had been reduced in status to a mathematical curiosity, a mere intellectual diversion.

7

Luoshu Miscellanea

> Mathematicians continue to make fascinating mathematical discoveries concerning magic squares, thus indicating that the luoshu *may still possess some unrevealed secrets. Many interesting puzzles have been created that rely on the magic square of order three. In addition, the* luoshu *and related magic squares have formed the basis for musical compositions, intriguing visual patterns, and* taiji *movements.*

Some Mathematical Considerations

Under the basic definition of a magic square (a square number array in which the sum of each row, column, and diagonal is the same), a one-celled square could exist, in theory. Of course, any square consisting of one number would be a magic square, but a trivial and uninteresting one. Richard Webster in his book *Talisman Magic* notes that medieval Christians employed a one-celled magic square as a symbol for God, while a magic square of order two, being impossible, they associated with the devil.[1] We can easily demonstrate that a magic square of order two is impossible. Assume that such a square exists. Let its elements be represented by the letters a, b, c, d, where each letter represents a different numerical value. Thus the square is of the form:

FIGURE 7.1

Let the magic sum for each row, column, and diagonal be k. So a + b = k, a + c = k, and so on. Solving just two of the possible six linear equations, simultaneously, we find:

$$a + b = k$$
$$-(a + c = k)$$
$$\overline{b - c = 0 \Rightarrow b = c}$$

This indicates that two numerical values are the same and that contradicts our definition of a magic square.

The smallest magic square of mathematical substance appears to be the square of order three. Early mathematical enthusiasts would also have come to this conclusion because they constructed natural squares using the counting numbers. Theon of Smyrna (ca. 130), a Neo-Pythagorean, has, at times, been credited with devising the first magic square in his work on arithmetic and astronomy known by its Latin name, *Expositio*; however, an examination of Theon's square reveals it is merely a natural square.[2] See figure 7.2.

Let us examine the structure of a third-order magic square. Again, we will represent the numerical elements of the square by nine letters of the alphabet, a through i. See figure 7.3.

1	4	7
2	5	8
3	6	9

FIGURE 7.2

a	b	c
d	e	f
g	h	i

FIGURE 7.3

Assume the sum of all the numbers in the square is m, then the magic constant for the square, k, will be m/3. Now consider the sum of the two diagonal and middle column numbers:

$$(a + e + i) + (g + e + c) + (b + e + h) = k + k + k = 3k = m$$

so $(a + e + i) + (g + e + c) + (b + e + h) = m$

rewriting the left side of this equation, we find:

$$(a + b + c) + 3e + (i + g + h) = m$$

since $(a + b + c) = (i + g + h) = k$

$$m/3 + 3e + m/3 = m \Rightarrow 3e = m/3$$

or $e = m/9$, e is the average value for the sequence and $3e = k$.

We have just shown that for the third-order magic square, three times the center number gives the magic constant. The principle can be generalized for any odd-order magic square.

Consider now the sum of the two diagonals' elements, $(a + e + i) + (g + e + c) = 2k$, rewriting the left side of the equation, we arrive at $(a + i) + 2e + (g + c) = 2k$. Since $e = k/3$, we have $(a + i) = (g + c) = 4k/3 = 4e$ but $(a + i) = (g + c)$ so $(a + i) = 2e$ and $(g + c) = 2e$.

Let $a = e - p$ and $g = e - q$ where p and q represent numbers of different values. Under this substitution, we have

$$(e - p + i) = 2e \Rightarrow i = e + p$$
$$(e - q + c) = 2e \Rightarrow c = e + q$$

Substituting these values back into the original magic square configuration, we have

e − p	b	e + q
d	e	f
e − q	b	e + p

FIGURE 7.4

Since this square is magic, the sum of the elements in the first column must equal the magic constant k, so:

$$(e - p) + d + (e - q) = 3e \Rightarrow d - q - p = e$$

or $d = (e + p + q)$.

By similar reasoning and computation, we can arrive at values for the remaining unknowns:

$$b = (e + p - q)$$
$$f = (e - p - q)$$
$$h = (e - p + q)$$

Thus, a third-order magic square can be constructed using three numbers e, p, q where $p \neq q$.[3]

e − p	e + p − q	e + q
e + p + q	e	e − p − q
e − q	e − p + q	e + p

FIGURE 7.5

The structure of this square readily reveals two properties of a third-order magic square: the value of the central entry is embedded in all other entries—a property that impressed the Chinese users of the *luoshu*—and the sum of the border entries, if taken diagonally pairwise, will equal twice the central entry, then the square is associated.

To construct a third-order normal magic square, that is, a magic square composed of the consecutive numbers 1 through 9: First, figure the magic sum. Since $1 + 2 + 3 + \ldots + 8 + 9 = 45$, and 45 divided by 3 (the order of square) equals 15, then the magic sum is 15. The central entry is 5 because 45 divided by 9 equals 5. After putting a 5 in the central cell, the remaining numbers can be filled in, starting with 1. From experience with a normal square, a seemingly natural position for 1 would be a corner cell. See figure 7.6. If this is done, a 9 must be placed in the opposing

	*	1
	5	*
9		

FIGURE 7.6

corner. When this happens, it prevents the numbers 8, 7, and 6 from occupying cells in the same row or column as 9 and dictates that the three numbers must occupy two possible cells (marked by an asterisk in figure 7.6), an impossible situation. Thus, 1 cannot occupy a corner cell, and when placed in a central row or column, its companion entry in the diagonally opposite cell must be 9. See figure 7.7.

	9	
	5	
	1	

FIGURE 7.7

At this point it becomes clear that the other numbers in 9's row must be 4 and 2. Putting the remaining numbers in the available cells results in the *luoshu*. This process reveals that the *luoshu* is truly unique.

The first method derived above will yield the *luoshu* when the values for p and q are selected from the number sets {1, 3} and {–1, –3}, respectively. The eight possible number combinations will result in the eight equivalent forms of the *luoshu* with p = 1, q = –3 supplying the standard form. The numerical entries form three arithmetical progressions:

(1) $e - p - q$, $\quad e - p$, $\quad e - p + q$
(2) $e - q$, $\quad e$, $\quad e + q$
(3) $e + p - q$, $\quad e + p$, $\quad e + p + q$

Within each progression, successive terms differ by the constant q; corresponding terms across progressions differ by the constant p. Adding the same amount to each element of a magic square, or multiplying each by the same constant, will result in another magic square—the progressive structure is preserved and other third-order magic squares result. Thus, it is possible to construct an infinite number of augmented third-order magic squares.

There are still other magic sums that can be derived from the *luoshu*, for example, the sum of the squares of the numbers in the first row equals the sum of the squares for the numbers in the third row. A similar relationship exists for the numbers in the first column and those in the third. In the 1970 edition of the *Mathematical Gazette*, R. Holmes of London discovered still another magic property.[4] If the digits of the rows, columns or diagonals, including broken diagonals, are treated as three-digit numbers, these numbers squared and added together, and the process repeated with the digits of the individual three-place numbers reversed, the same result will be obtained. See figure 7.8.

4	9	2
3	5	7
8	1	6

rows: $(492^2 + 357^2 + 816^2) = (294^2 + 753^2 + 618^2)$
columns: $(438^2 + 951^2 + 276^2) = (834^2 + 159^2 + 672^2)$
diagonals: $(456^2 + 978^2 + 231^2) = (654^2 + 879^2 + 132^2)$
          $(258^2 + 714^2 + 936^2) = (852^2 + 417^2 + 639^2)$

FIGURE 7.8

Further, these identities still hold when the middle digit or any two corresponding digits of the six addends are deleted.

In 1975, Hwa Suk Hahn of Carollton, Georgia found that the sum of row products equaled the sum of column products in the *luoshu*:

$[(4 \times 9 \times 2) + (3 \times 5 \times 7) + (8 \times 1 \times 6)] = [(4 \times 3 \times 8) + (9 \times 5 \times 1) + (2 \times 7 \times 6)]$

This property holds for all third-order magic squares. Hahn termed magic squares with this property "balanced squares."[5] More recently, Martin Gardner, wondering if under certain conditions this property could also be extended to include the product-sum of the diagonals, sought out the mathematical assistance of a number theorist friend, John Robertson.[6] Indeed, Robertson found such squares and proved that an infinite number of them existed. Three such "Robertson squares" are shown in figure 7.9 with their magic product-sums given.

33	2	43
36	26	16
9	50	19

2 6364

41	4	57
50	34	18
11	64	27

5 8956

67	2	81
64	50	36
19	98	33

1 8 7500

FIGURE 7.9

Further, John Robertson found a principle upon which such squares can be constructed. Choose three numbers: x, y, and z, whose squares form an arithmetic progression. Then using the analytic formula forms for a third-order magic square as given above: let e = 2y, p = x and q = z. For the squares shown in figure 7.9, these values are respectively: 26, 7, 17; 34, 7, 23; and 50, 17, 31. Robertson also found that any set of Pythagorean triples, that is, numbers, a, b, and c, that satisfy the Pythagorean relationship $a^2 + b^2 = c^2$, gives rise to a "Robertson square."[7] A unique feature of the *luoshu* is that the product-sums of its rows and columns equals its magic constant squared, thus, $15^2 = 225$.

Many problem-solvers who have come into contact with magic squares and who have knowledge of higher mathematics are often struck by their physical similarity to matrices. A matrix (the singular form of *matrices*) is a rectangular array of numbers that represent certain mathematical relationships possible within a mathematical structure called a vector space. Some matrices are square configurations of numbers just like magic squares. Matrices, as members of vector space, satisfy certain specific conditions: first, there are two operations defined on them, an "addition process" and a "multiplication process": second, these operations must possess certain properties that define a vector space.[8] Treating two *luoshu*-derived magic squares as matrices, let us add them. The addition is performed by taking the sums of all corresponding cells of the squares: that is, the two upper righthand corner entries are added together; the two upper middle entries are added together, and so on as shown in figure 7.10.

8	18	4
6	10	14
16	2	12

+

6	21	18
27	15	3
12	9	24

=

14	39	22
33	25	17
28	11	36

FIGURE 7.10

The result is another magic square. This is impressive—the magic square property is preserved under the operation of matrix addition. However, a little further experimentation reveals a problem with our blossoming theory: such additions will produce squares with repeated numbers, violating one of the defining conditions imposed in our discussion of magic squares, namely, that no elements of the square may resulted in a wide and varied array of mathematical creations. Ann-Lee Wang, writing in the 1995 issue of *Mathematics in School*, describes the concept of hollow magic squares.[17] The theory is prefaced by an old, possibly apocryphal, Chinese legend of a commander defending a square walled city with only a handful of soldiers. As Wang relates: "The commander devised a plan wherein the soldiers were deployed at intervals on the wall in various patterns. The spies of the enemy force observed that although the number of soldiers at any particular spot on the wall kept changing, the total number along any wall remained the same."[18] The enemy commander believed that his opponent was using magic against him and he withdrew—the city was saved. This defensive strategy was based on devising a "hollow magic square," that is, a square number configuration forming a border where the rows and columns have the same sum. Using the numbers 1 to 8, six third-order hollow magic squares are possible. Squares with magic constants 12 and 13 are shown in figure 7.12.

6	5	1
4		8
2	7	3

5	2	6
7		3
1	8	4

FIGURE 7.12

An antimagic square of order n is a square comprised of the positive integers $1\text{-}n^2$ such that no two rows, columns, or diagonals have the same sum. It appears that the mathematical puzzle master Sam Lloyd was the first to consider such squares in the late nineteenth century.[19]

An antimagic square of order three is given in figure 7.13.

1	2	3
8	9	4
7	6	5

FIGURE 7.13

Further, John Robertson found a principle upon which such squares can be constructed. Choose three numbers: x, y, and z, whose squares form an arithmetic progression. Then using the analytic formula forms for a third-order magic square as given above: let $e = 2y$, $p = x$ and $q = z$. For the squares shown in figure 7.9, these values are respectively: 26, 7, 17; 34, 7, 23; and 50, 17, 31. Robertson also found that any set of Pythagorean triples, that is, numbers, a, b, and c, that satisfy the Pythagorean relationship $a^2 + b^2 = c^2$, gives rise to a "Robertson square."[7] A unique feature of the *luoshu* is that the product-sums of its rows and columns equals its magic constant squared, thus, $15^2 = 225$.

Many problem-solvers who have come into contact with magic squares and who have knowledge of higher mathematics are often struck by their physical similarity to matrices. A matrix (the singular form of *matrices*) is a rectangular array of numbers that represent certain mathematical relationships possible within a mathematical structure called a vector space. Some matrices are square configurations of numbers just like magic squares. Matrices, as members of vector space, satisfy certain specific conditions: first, there are two operations defined on them, an "addition process" and a "multiplication process": second, these operations must possess certain properties that define a vector space.[8] Treating two *luoshu*-derived magic squares as matrices, let us add them. The addition is performed by taking the sums of all corresponding cells of the squares: that is, the two upper righthand corner entries are added together; the two upper middle entries are added together, and so on as shown in figure 7.10.

8	18	4
6	10	14
16	2	12

+

6	21	18
27	15	3
12	9	24

=

14	39	22
33	25	17
28	11	36

FIGURE 7.10

The result is another magic square. This is impressive—the magic square property is preserved under the operation of matrix addition. However, a little further experimentation reveals a problem with our blossoming theory: such additions will produce squares with repeated numbers, violating one of the defining conditions imposed in our discussion of magic squares, namely, that no elements of the square may be repeated. By eliminating this stipulation and defining a magic square as a square array of numbers where the sum of each row, column, and diagonal is the same, magic squares form a vector space.[9] It seems that the mathematical community at large has adopted this definition of a magic square.[10]

Combining *luoshu* magic squares under the operation of matrix multiplication yields interesting results.[11] A *luoshu* square multiplied by a *luoshu* square produces a semi-magic square with repeating elements whose magic constant is 225, or 15^2.[12] If this resulting square is again multiplied by the *luoshu* square, so that in essence we have the *luoshu* cubed or raised to the same power as its order, the product found is a "true" magic square whose magic constant is 3,375 or $(15)^3$. This square is shown in figure 7.11.

1149	1029	1197
1173	1125	1077
1053	1221	1101

FIGURE 7.11

If magic squares are investigated as elements of a vector space, a wide horizon of research opportunities and problem challenges emerge.[13]

Luoshu Puzzles

Once the magic square of order three entered the domain of mathematical interest, it became the basis for mathematical problems and puzzles. A common way to use the *luoshu* magic square as a basis for creating a puzzle is to challenge someone to arrange the numbers 1 through 9 into a square so that all rows, columns, and diagonals add up to the same sum, that is, form the *luoshu*. A variation of this puzzle is found in the medieval monks' problem where three monks (brothers) must each draw a quantity of wine. The monks each have three containers they can fill and they obtain their wine from the contents of nine containers, which hold 1 to 9 measures of wine respectively. The question posed is, How might this task be accomplished so that all the monks end up with the same amount of wine? Each will have fifteen measures of wine, drawing in quantities prescribed by the rows, columns, or diagonal numbers of the *luoshu*.[14]

Puzzle books still challenge readers with situations such as:

Given the square of numbers:

4	3	2
7	1	9
6	5	8

can you arrange the nine digits in the square so that in all possible eight directions [diagonals included], the difference between one of the digits and the sum of the remaining two will always be the same? In the given square, it will be found that all the rows and columns give the difference 3, (thus (4 +2) − 3, (1 + 9) − 7, (6 + 5) − 8 etc.) but the two diagonals are wrong because 8 − (4 + 1) and 6 − (1 + 2) is not allowed: the sum of the two must not be taken from the single digit, but the single digit from the sum. How many solutions are there?[15]

It is found that the *luoshu* and its variations supply the required solutions. A particularly interesting problem involving magic squares of order three appeared in the problem section of the 1947 issue of the *American Mathematical Monthly*. It goes as follows:

> The court mathematician once received his salary for a year's service all at one time, and all in silver "dollars," which he proceeded to arrange in nine unequal piles, making a magic square. The king looked, and admired, but complained that there was not a single prime number in any of the piles. "If I had but nine coins more," said the mathematician, "I could add one coin to each pile and make a magic square with every number prime." They investigated and found that this was indeed true. The king was about to give him nine more dollars, when the court jester said "Wait!" Then the jester subtracted one coin from each pile instead; and they found in this case also a magic square with every element a prime number. The jester kept the nine "dollars." How much salary must the mathematician have been receiving?[16]

Mathematicians are a creative and curious bunch who often like to push a concept to its limits with considerations of "What if . . . ? and "Suppose . . . ?" Such considerations, when applied to the *luoshu* and the more general conept of the magic square of order three have resulted in a wide and varied array of mathematical creations. Ann-Lee Wang, writing in the 1995 issue of *Mathematics in School*, describes the concept of hollow magic squares.[17] The theory is prefaced by an old, possibly apocryphal, Chinese legend of a commander defending a square walled city with only a handful of soldiers. As Wang relates: "The commander devised a plan wherein the soldiers were deployed at intervals on the wall in various patterns. The spies of the enemy force observed that although the number of soldiers at any particular spot on the wall kept changing, the total number along any wall remained the same."[18] The enemy commander believed that his opponent was using magic against him and he withdrew—the city was saved. This defensive strategy was based on devising a "hollow magic square," that is, a square number configuration forming a border where the rows and columns have the same sum. Using the numbers 1 to 8, six third-order hollow magic squares are possible. Squares with magic constants 12 and 13 are shown in figure 7.12.

6	5	1
4		8
2	7	3

5	2	6
7		3
1	8	4

FIGURE 7.12

An antimagic square of order n is a square comprised of the positive integers $1\text{-}n^2$ such that no two rows, columns, or diagonals have the same sum. It appears that the mathematical puzzle master Sam Lloyd was the first to consider such squares in the late nineteenth century.[19]

An antimagic square of order three is given in figure 7.13.

1	2	3
8	9	4
7	6	5

FIGURE 7.13

In 1951, Dewey Duncan extended this concept a step further by defining a heterosquare as an antimagic square in which the sum of all broken diagonals also differed. It has been proven that no heterosquares of order three exist.[20]

The most straightforward magic square challenge is to specify that certain numbers be used or specific magic sums obtained. Several such squares are shown in figure 7.14:

- (a) comprised of odd integers
- (b) comprised of even integers
- (c) comprised of primes
- (d) comprised of consecutive composite numbers
- (e) comprised of consecutive primes.[21]

15	1	11
5	9	13
7	17	3

(a)

8	18	4
6	10	14
16	2	12

(b)

101	5	71
29	59	89
47	113	17

(c)

117	116	121
122	118	114
115	120	119

(d)

1 480 028 201	1 480 028 129	1 480 028 183
1 480 028 153	1 480 028 171	1 480 028 189
1 480 028 159	1 480 028 213	1 480 028 141

(e)

FIGURE 7.14

In 1987, Martin Gardner offered a prize of $100 to anyone who could construct a third-order magic square using consecutive prime numbers. Harry Nelson of the Lawrence Livermore Laboratories, assisted by a Cray computer, arrived at a solution square given in figure 7.14(e). In 1996, Gardner renewed the same offer but this time it was for the construction of a third-order magic square in which all nine numbers were distinct squares.[22] The resolution of this problem rests with finding a solution for a particular elliptic curve of the form $y^2 = x^3 - n^2x$, a similar problem to that of Fermat's Last Theorem.[23] To date, no one has claimed this prize. An intriguing third-order magic square is the palindromic square of the "Beast." Each number in the square gives the same value read forward or backwards and the magic constant is 666, the number of the Beast in the medieval system of gematria. This square is shown in figure 7.15.[24]

232	313	121
111	222	333
323	131	212

FIGURE 7.15

Historically, the magical property of a magic square arises from the concept of a constant sum obtained by adding the numbers in the rows, columns, and diagonals. It results from the use of addition but magic squares can also be defined based on the operations of subtraction, multiplication, and division. Figure 7.16 provides examples of: (a) a multiplication magic square whose product along any row, column, or main diagonal is 216 and (b) a division magic square where if the extreme numbers of a row, column, or diagonal are multiplied together and the resulting product divided by the middle number of the line, the constant 6 will appear.[25]

12	1	18
9	6	4
2	36	3

(a)

3	1	2
9	6	4
18	36	12

(b)

FIGURE 7.16

A magic square is a two-dimensional configuration. It lies in the plane of its construction. Mathematically speaking, a natural extension of a magic square would be a three-dimensional analogue, a magic cube.[26] Figure 7.17 shows a magic cube where the three numbers comprising rows, columns, or diagonal lines which pass through the cube and contain the central entry, 14, will produce the magic constant 42.

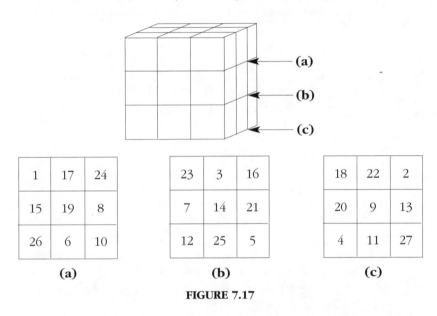

FIGURE 7.17

Note that the number squares that form the individual layers of the cube are not magic squares in themselves. One of the most unusual descendants of the *luoshu* is the alphamagic square discovered by Lee Sallows in 1985.[27] Sallows is a British engineer and recreational mathematics enthusiast employed by the University of Nijmegen in the Netherlands. The concept of an alphamagic square is best explained through the use of an illustration. Figure 7.18 shows two magic squares.

5	22	18
28	15	2
12	8	25

(a)

4	9	8
11	7	3
6	5	10

(b)

FIGURE 7.18

If one takes the English language number word for each entry shown in the square at the left, counts the number of letters in that word and places that number in the corresponding empty cell of a square of the same dimension, the square at the right results. For example, 5 is the entry in the left upper corner cell of the left square; the word "five" contains 4 letters; thus, the number 4 is placed in the upper lefthand corner cell of the right square. The square on the right is the alphamagic image of its companion on the left. Sallows has produced such alphamagic squares using a variety of modern languages.[28]

Another Time, Another Place, Another Legend

In 1887 in England, a privately published book entitled *The Origin of Tree Worship* describe Druidical practices of the Celtic and Germanic yew cult in pre-Christian Europe. The book relied on medieval source material, long believed to have been lost. Only six copies of the work were printed, one of which eventually found its way into the British Museum, but by 1888 that volume had been misplaced. The remaining five books have also disappeared. Included in the Celtic lore in the book is the story of King Mi's pilgrimage to a sacred grove of trees in *Eohdalir*, Valley of the Yew, where he performed a sacred ceremony that involved carving a runic charm on the trunk of the hallowed *Li*, oldest of the standing yews.

In 1985, *The Origin of Tree Worship* reappeared in the holdings of the British Museum. Runic script is an ancient angular alphabetic script, consisting of characters called runes and employed by mostly Scandinavian tribes to inscribe magico-ritualistic messages. Sallows, an amateur runologist, was permitted to view the find. He translated it as a series of number words and noted that the number of runes per line, twenty-two, corresponded exactly to the number of letters required to render the line into English. Sallows also noted that the indicated numbers formed a magic square, which he christened the *Li shu* in honor of the hallowed *Li* but also reflecting on the *luoshu*. Fascinated with his rune-letter correspondence, he formed a square of numbers based on the number of letters in the number words; the result of this effort was the alphamagic square shown in figure 7.18(b). The runic charm and its translation are given in figure 7.19.[29]

5	22	18
28	15	2
12	8	25

FIGURE 7.19

Mr. Browne's Illustrious Magic Square

In an early twentieth-century edition of *The Monist*, an article appeared entitled "Magics and Pythagorean Numbers" by Mr. C. A. Browne, Jr. Mr. Browne was versed in neo-Pythagorean number lore and an astute creator of magic squares. The article later found its way into the book, *Magic Squares and Cubes*, first published in 1917.[30]

In the article, Browne examines the origins of certain significant Pythagorean numbers, particularly the number 729. In Plato's *Republic*, 729 represents the difference between a "kingly man" and a tyrant as well as the "number of the state." Building upon his knowledge of ancient literature and numerology, Brown evolves an admittedly fanciful theory that supplied answers and produced an interesting magic square. According to Brown, the proportions of the Pythagorean *Tetractys* were embedded into many old number schemes. Plutarch assigned numbers to the planets using such a scheme: 729, (3^6), represented the Sun; 243, (3^5), Venus; 81, (3^4), Mercury; 27, (3^3), the Moon; 9, (3^2), Earth; and 3, (3^1), Antichathon, an imaginary anti-Earth. Plato, in his *Timaeus,* combined the proportions into one series: 1, 2, 3, 4, 9, 8, 27. In Pythagorean thinking, the number 27 dominates Plato's series because it is the sum of its predecessors: $1 + 2 + 3 + 4 + 9 + 8 = 27$. Browne notes the significance of the number 27, particularly its association with the Moon—the Moon completes its elliptical orbit about the Earth in 27 days. Further, 729 = 27 × 27, twenty-seven squared. Buttressed by this knowledge, he seeks to unravel his number mysteries by the use of a 27 × 27 normal magic square comprised of the numbers 1 to 729. The square he devised is shown in figure 7.20. Its method of construction appears to be unique for its time.

Legacy of the Luoshu

352	381	326	439	468	413	274	303	248	613	642	587	700	729	674	535	564	509	118	147	92	205	234	179	40	69	14
327	353	379	414	440	466	249	275	301	588	614	640	675	701	727	510	536	562	93	119	145	180	206	232	15	41	67
380	325	354	467	412	441	302	247	276	641	586	615	728	673	702	563	508	537	146	91	120	233	178	207	68	13	42
277	306	251	355	384	329	433	462	407	538	567	512	616	645	590	694	723	668	43	72	17	121	150	95	199	228	173
252	278	304	330	356	382	408	434	460	513	539	565	591	617	643	669	695	721	18	44	70	96	122	148	174	200	226
305	250	279	383	328	357	461	406	435	566	511	540	644	589	618	722	667	696	71	16	45	149	94	123	227	172	201
436	465	410	271	300	245	358	387	332	697	726	671	532	561	606	619	648	593	202	231	176	37	66	11	124	153	98
411	437	463	246	272	298	333	359	385	672	698	724	507	533	559	594	620	646	177	203	229	12	38	64	99	125	151
464	409	438	299	244	273	386	331	360	725	670	699	560	505	534	647	592	621	230	175	204	65	10	39	152	97	126
127	156	101	214	243	188	49	78	23	361	390	335	448	477	422	283	312	257	595	624	569	682	711	656	517	546	491
102	128	154	189	215	241	24	50	76	336	362	388	423	449	475	258	284	310	570	596	622	657	683	709	492	518	544
155	100	129	242	187	216	77	22	51	389	334	363	476	421	450	311	256	285	623	568	597	710	655	684	545	490	519
52	81	26	130	159	104	208	237	182	286	315	260	364	393	338	442	471	416	520	549	494	598	627	572	676	705	650
27	53	79	105	131	157	183	209	235	261	287	313	339	365	391	417	443	469	495	521	547	573	599	625	651	677	703
80	25	54	158	103	132	236	181	210	314	259	288	392	337	366	470	415	444	548	493	522	626	571	600	704	649	678
211	240	185	46	75	20	133	162	107	445	474	419	280	309	254	367	396	341	679	708	653	514	543	488	601	630	575
186	212	238	21	47	73	108	134	160	420	446	472	255	281	307	342	368	394	654	680	706	489	515	541	576	602	628
239	184	213	74	19	48	161	106	135	473	418	447	308	253	282	395	340	369	707	652	681	542	487	516	629	574	603
604	633	578	691	720	665	526	555	500	109	138	83	196	225	170	31	60	5	370	399	344	457	486	431	292	321	266
579	605	631	666	692	718	501	527	553	84	110	136	171	197	223	6	32	58	345	371	397	432	458	484	267	293	319
632	577	606	719	664	693	554	499	528	137	82	111	224	169	198	59	4	33	398	343	372	485	430	459	320	265	294
529	558	503	607	636	581	685	714	659	34	63	8	112	141	86	190	219	164	295	324	269	373	402	347	451	480	425
504	530	556	582	608	634	660	686	712	9	35	61	87	113	139	165	191	217	270	296	322	348	374	400	426	452	478
557	502	531	635	580	609	713	658	687	62	7	36	140	85	114	218	163	192	323	268	297	401	346	375	479	424	453
688	717	662	523	552	497	610	639	584	193	222	167	28	57	2	115	144	89	454	483	428	289	318	263	376	405	350
663	689	715	498	524	550	585	611	637	168	194	220	3	29	55	90	116	142	429	455	481	264	290	316	351	377	403
716	661	690	551	496	525	638	583	612	221	166	195	56	1	30	143	88	117	482	427	456	317	262	291	404	349	378

FIGURE 7.20

Browne recognized that he could construct the *luoshu* square of order three by following the *yubu* path that Daoist priests followed. He then repeated the process over and over again in slightly different numerical contexts until he arrived at his twenty-seventh order square. The *yubu* path consists of three separate movements: an "upper wedge," a "slash down to the right," and a "lower wedge." These are illustrated in figure 7.21.

Suppose now we wish to construct a ninth-order magic square using the *yubu* method. This can be done by starting with a natural number square for the numbers 1 to 81 written according to a Chinese lexicographical ordering. See figure 7.22.

The square is then partitioned into nine three-by-three number squares; let us designate them 1 to 9 in the same lexicographical order proceeding from the top to bottom, right to left. See figure 7.23.

Luoshu *Miscellanea*

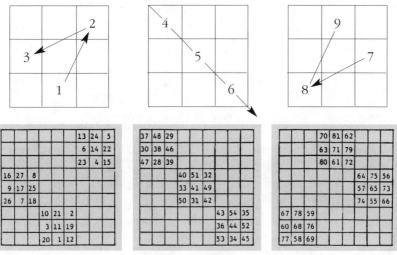

FIGURE 7.21

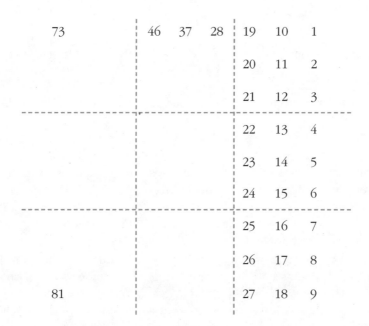

FIGURE 7.22

FIGURE 7.23

Each three-by-three number square is now made into a magic square by applying the *yubu* technique. The resulting nine magic squares are then positioned into the matrix of a ninth-order magic square again using a *yubu* pattern as follows:

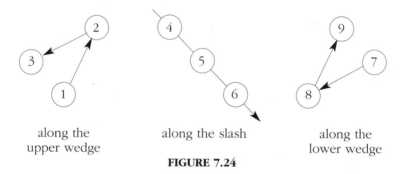

along the upper wedge   along the slash   along the lower wedge

FIGURE 7.24

The combined results yield a composite magic square whose magic constant is 369. See figure 7.25.

This square was developed by applying the *yubu* technique at two independent levels or stages: first, it was used to form the three-by-three magic squares and second, to order those magic squares to form the larger nine-by-nine magic square. If we began with a 27 × 27 natural number square, partitioned it into nine different nine-by-nine number squares, converted each one of those by *yubu* iteration into a magic square, and then applied the *yubu* technique again to order the resulting nine magic squares, we arrive at Browne's square as shown in figure 7.20. This square is the result of a three-stage process based on the *yubu* movement.

In his quest to solve the number mysteries, Browne had somehow to arrive at the specific numbers, certainly at least the kingly number 729, and had to meet the ancient conditions regarding the numbers. A hint as to those conditions is related in a conversation from the *Republic* between Socrates and fellow philosopher Glaucon:

37	48	29	70	81	62	13	24	5
30	38	46	63	71	79	6	14	22
47	28	39	80	61	72	23	4	15
16	27	8	40	51	32	64	75	56
9	17	25	33	41	49	57	65	73
26	7	18	50	31	42	74	55	66
67	78	59	10	21	2	43	54	35
60	68	76	3	11	19	36	44	52
77	58	69	20	1	12	53	34	45

FIGURE 7.25

Socrates: And if a person tells the measure of the interval which separates the king from the tyrant in truth of pleasure, he will find him, when the multiplication is completed, living 729 times more pleasantly, and the tyrant more painfully by this same interval.

Glaucon: What a wonderful calculation.

Socrates: Yet a true calculation and a number which closely concerns human life, if human life is concerned with days and nights and months and years.[31]

Thus, in the calculations from which the number 729 emerges there must be a chronological connection with days, months, and years. Incredibly, such a connection can be found in the large magic square: the central number is 365 (the number of days in a year); the most central three-by-three magic square has a magic sum of 1,095 (the number of days in three years); next, the central nine-by-nine magic square possesses the magic constant 3,285 (the number of days in nine years); and finally, the magic constant for the whole square is 9,855 (the number of days in 27 years). Viewed as a nested set of magic squares, the magic constants of the respective subsquares form a geometric progression: 1, 3, 9, 27. Delighted with his find, Browne shaded alternate cells in his square to emphasize the yearly connection: 365 white cells corresponded to days and 364 dark cells corresponded to nights. With some further justifications, Browne concluded that the magic constant for his unusual square, 9,855, was the elusive "number of the State."

Charles W. Trigg, an avid problem solver and student of magic squares, also constructed a 27 × 27 magic square by building upon the *luoshu*. Trigg took the *luoshu* and augmented it repeatedly by adding nine to all entries. He then combined the resultant nine third-order magic squares into a composite ninth-order square; 134, 217, 728 such squares were possible. These ninth-order squares could in turn be manipulated and then combined to form a composite twenty-seventh-order square. Trigg calculated that $(8^9)^2$ or 18, 014, 398, 509, 481, 984 such squares are possible. All of these squares are comprised of the first 729 integers and possess the magic constant 9,855.[32]

Feel the Rhythm

The symmetries of the *luoshu* and the patterns of its numbers have attracted attention from the earliest times. Daoist priests, in a dance of melodic shuffling, still trace out the path of *yubu* in some of their ceremonies.[33] In the West, since the time of Agrippa, magicians have also evoked the *yubu's* path as a source of empowerment. These are two cases of the rhythms embodied in the *luoshu* being incorporated in ritualistic movements: for Daoist priests, it takes the form of a dance, for Western seekers of magic, it is a path to be traced out. The rhythms of the *luoshu* have been used in other ways.

Sir Peter Maxwell Davies is one of Great Britian's most gifted avant-garde composers. Sir Peter blends seemingly incongruous elements into dramatic and harmonious masterpieces. One of these elements has been magic squares, or rather, the patterns Davies perceived in the workings of the squares. While his exact technique remains elusive, some association between specific works and particular magic squares has been documented.[34] In Davies's composition *Ave Maris Stella* (1975), an instrumental sextet to be played without a conductor, Davies took a traditional Ave Maris Stella plainsong form and "projected" it through the Magic Square of the Moon. This square is of order nine but by the technique of repeatedly adding together the digits for the number in each cell until the resulting square consists of only single digits from 1 to 9, a repeating pattern is discerned and lends itself to a system for governing durations throughout the piece. It is not clear if Davies has ever actually used the *luoshu* itself as a basis for one of his compositions but, in the realm of music, he has formulated a method that could be adapted to the rhythms of the magic square of order three.

As noted by Daoist priests, the *yubu* path denotes a rhythmic path through the *luoshu*. Claude Bragdon, a prominent architect and noted occultist during the first half of the twentieth century, built upon *luoshu* rhythms to produce interesting visual ornamental designs.[35] Four of his designs are presented in figure 7.26.[36] In his work, Bragdon referred to the paths within magic squares as "magic lines." In figure 7.27, he has used the *yubu* magic line to construct an appealing frieze pattern.[37] Bragdon experienced and communicated the beauty and the rhythms he saw in magic squares, viewing them as "a window opening into the world of the wondrous."[38]

FIGURE 7.26

Modern mathematics teachers have also appreciated the rhythms of the *luoshu* as an example of mathematical patterns and symmetries. They devised this simple exercise. They ask students to trace out the *yubu* circuit as if a rectangular grid were imposed on the *luoshu*.[39] In this way, students must associate each number with a definite point in space. After students complete the path and have the initial square of

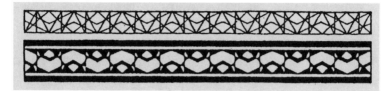

FIGURE 7.27

the *luoshu* partitioned into convex regions, they must color in the regions using the minimum number of colors necessary to distinguish adjacent regions—for the *luoshu*, two colors will suffice. Two variations on the resulting design are possible depending on whether there is a line drawn between the initial 1 and the final 9. Both variations are shown in figure 7.28.

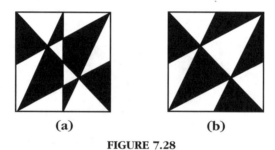

FIGURE 7.28

If one then treats the completed square like a tile and combines it with three variations of itself—created by rotation or reflection—to form a larger square, a pleasing mosaic effect is achieved that can be repeated to tile the complete plane. See figure 7.29, where two such patterns are developed that employ the basic design given in figure 7.28(b).

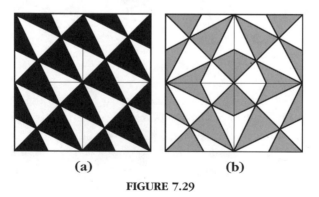

FIGURE 7.29

Luoshu *Miscellanea*

With the power of the modern electronic computer, even more complex and interesting visual rhythms can be created from the *luoshu*.[40] Using black-and-white color coding plus iteration, the interplay of *yin* and *yang* can be explored. If we begin by coloring the cells of the corner *yin* numbers black and the remaining cells for the *yang* numbers white, a cruciform image emerges from the *luoshu*. We can then generate larger and larger squares by combining many *luoshu* squares into more expansive patterns: a nine-by-nine square formed from nine *luoshu* squares; a 27 × 27 square formed out of twenty-seven *luoshu*s, and so on (3, 3^2, 3^3, . . .). Throughout these iterations, one principle will be followed: the four corner cells will have the opposite coloring of the five inner cells. A one-stage iteration, moving from the *luoshu* to the composite ninth-order *luoshu*, is shown in figure 7.30.

FIGURE 7.30

A large, elaborate cruciform image emerges, growing from the center of the square and emanating outward to the centers of the four sides. This is depicted in figure 7.31.[41]

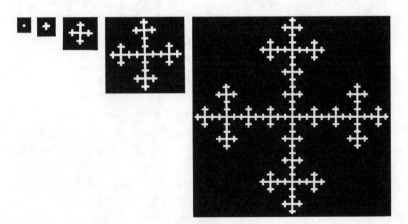

FIGURE 7.31

144 *Legacy of the* Luoshu

An 81 × 81 *luoshu*-generated square with two-color coding is shown in figure 7.32. Note the central cross.

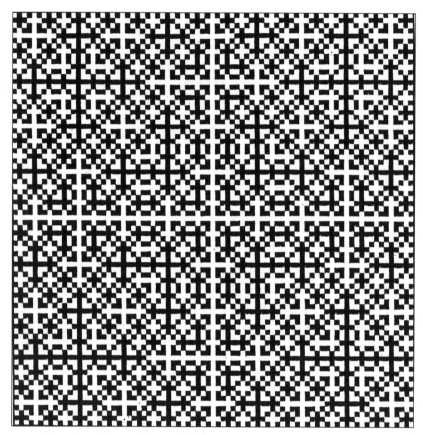

FIGURE 7.32

Another technique for exploring hidden rhythms within the *luoshu* is to first alter the numbers to find a mathematical pattern, then develop and extend the pattern, and finally, turn it into a visual display. Let us use such a technique on Browne's illustrious magic square (shown in figure 7.20 above). If we take the number in each cell and add its digits together, several times if needed, until we arrive at a single digit number, we convert the whole square to a square in which the numbers vary from 1 to 9. When this is done, the square becomes a composite square of nine identical nine-by-nine squares. One of these subsquares is shown in figure 7.33. The nine-by-nine square is itself a composite square

1	3	2	7	2	9	4	6	5
3	2	1	9	8	7	6	5	4
2	1	3	5	4	6	5	4	6
7	9	8	4	6	5	1	3	2
9	8	7	6	5	4	3	2	1
8	7	9	5	4	6	2	1	3
4	6	5	1	3	2	7	9	8
6	5	4	3	2	1	9	8	7
5	4	6	2	1	3	8	7	9

FIGURE 7.33

formed with nine three-by-three number squares, each displaying a set of permutations on three consecutive numbers within the natural sequence 1, 2, 3 . . ., 9. If we look at the ordering of these three-by-three squares within the complex of Browne's whole square, several cyclic patterns emerge: left to right; top to bottom; and diagonally from the upper right to the lower left. If these cycles are viewed as continuous, the magic square looks like the surface of a torus. Using a two-color shading scheme with black covering even-number cells and white covering odd-number cells, Browne's magic square reveals its visual rhythms (see figure 7.34).

Beholders of the *luoshu* and related magic squares have perceived many things in them, not the least of which are intriguing visual patterns and spatial relationships.

Taijiquan, the *Luoshu*, and Immortality

As early as the fourth century B.C.E., Chinese shamans, in pursuit of material immortality, were manipulating substances and concocting potions deemed to prolong life. Alchemy in China arose from these early efforts to find "elixirs of immortality." Daoism absorbed this quest

FIGURE 7.34

and refocused it on duplicating the processes of nature which resisted corruption and death. For example, gold is impervious to decay; therefore, in the theory of a Daoist alchemist, if gold was consumed as an ingredient of an elixir, it was thought to assist its recipient in achieving longevity. Daoist sages who could produce elixirs of immortality were accorded the honorary title of *hsien,* "immortal."

Zhang Sanfeng (ca. 1314–1417) was one such honorary immortal. A Daoist monk, he was widely recognized for his ability as an alchemist. Legend relates that one morning Zhang was awakened by the sounds of a struggle beneath his window. Investigating, he found a crane locked in deadly combat with a snake. Noting the combatants' series of thrusts and parries, the monk envisioned a method of physical exercise and self-conditioning based on a balanced alternation of applying force and yielding, attacking and evading—analogous to the concept of *yinyang.* He further developed his theory by observing the bodily dynamics of other wild creatures and the movements of nature as expressed in the flowing of water, the blowing of wind, the bending of trees and grass, and the passage of clouds across the sky. Zhang choreographed these natural movements into a regimen of sparring exercises, *taijiquan* [Supreme Ultimate Boxing], commonly referred to as *"taiji"* (or t'ai chi). *Taiji* movements, performed in a state of meditation and coordinated with breathing, are flowing, rhythmical, cyclical, and sequential. They have been described as "swimming in air" and a "ritual dance of meditation." Through this series of exercises, an individual focuses the flow of *qi* within his or her body and balances *yinyang* forces to achieve an inner harmony and state of well-being. Zhang Sanfeng believed that the attainment of this internal equilibrium resulted in perfect health and thus the achievement of longevity.

Taiji sparring movements, when performed barehanded against an opponent, or with a lance or sword, are a form of martial arts. As *taiji* techniques and practices evolved, they absorbed other Daoist mystical beliefs including those involving the *luoshu*. *Luoshu* dynamics provided patterns for coordination and emulation. Systems of steps, fistic forms, and methods of bodily motion were derived from *luoshu* theories. Many *taiji* masters felt that employing the *luoshu* as a training guide led to the very threshold of immortality. Master Sun Lutang (1860–1933) taught his students that:

> Through the practice of *taijiquan,* the yang is nourished and brought forth. Through the practice of *taiji* lance, the yin essence is aroused and enfolds. Through the union of the yang with yin, the five forces of water, wood, earth, fire, and gold come into being and merge within and throughout the body and lance. Through the *Luo-shu,* one embarks to be inscribed upon the Register of the Immortals.[42]

In the 1930s a brief thesis on *taijiquan* was published in Shanghai. It described the secret nature of *luoshu* applications as revered by the Chen clan, practitioners of their own form of *taiji,* and several other important schools of the discipline:

> The *Luo-shu*, according to many Chen family elders, is composed of both an inner and outer form, both halves embracing a myriad of meanings. The inner, in relation to the human physique, refers to the inner movement and location of *qi* in movement. The outer refers to a number of specific areas, some of which are: standing practice—in preparation for movement; combat—areas to strike and guard; and weapons—areas to focus one's attack for maiming and overcoming the opponent using lance, sword, knife, and so forth . . .

The text's author, Chen Weiming further noted:

These methods are especially taught by some of the elder teachers of the Chen *taiji* boxing style, Yang Luchan's original version of the Yang family art, and the Sun Lutang school. My master, Yang Chengfu, said that Yang Luchan learned both the inner and outer *Luo-shu* applications and thus was able to attain supreme boxing skills. He passed on this knowledge to his sons, but it was Yang Jianhou who most ardently practiced his boxing according to his ever-deepening understanding of the *Luo-shu* though it was Yang Banhou, in his later years, who began teaching these guarded methods outside the Yang family circle.[43]

While *luoshu* techniques were reserved as secret training for special students, some fragments of information are available. A solo boxing posture is demonstrated in figure 7.35(a) where the alignment of the numbers 8, 5, 2 is stressed, the right-diagonal of the *luoshu*. This illustration seems to demonstrate the *taiji* principle that all movements are "initiated from the legs, controlled by the waist, and shaped by the hands and fingers."[44] In the posture depicted, the anchoring foot is *yang* (8) which combines with the balance and control of the waist, *yin* (5) which, in turn, unites with the thrust of the hand, *yang*, (2), to achieve fulfillment in 15. A paired-lance encounter demonstrated in figure 7.35(b) is less obvious in its *luoshu* interpretation; however, a parry and pivot at the center 5 allows the left combatant to swing the point of his lance backwards and upwards, achieving a 9, 5, 1 alignment before thrusting at the throat of his opponent in an 8, 5, 2 movement.

In "saber and sword" practice, the blades' angles reflect on the *luoshu* arrangement. See figure 7.35 (c). In an instance of personal combat, when two individuals touch, it is believed that the *luoshu* unites between them and governs the contest. Changes in the *luoshu* path to victory depend on the proponents' postures, yielding moves, and the direction of force at specific points. See figure 7.35 (d). A successful yielding or advancing movement results in attaining the *luoshu* sum of 15.

The *luoshu* has influenced every aspect of both the Sun and Chen styles of *taijiquan*.

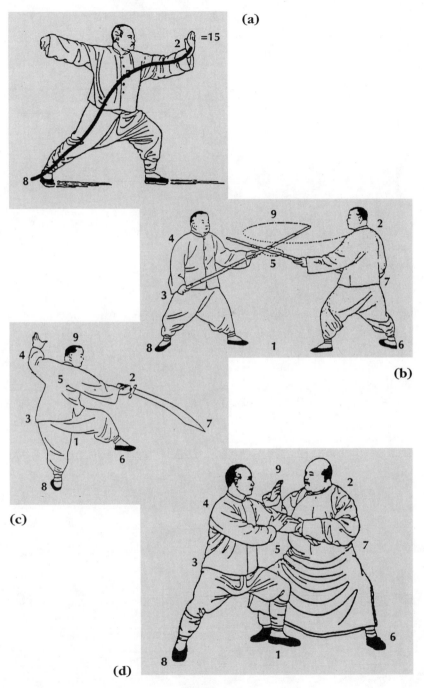

FIGURE 7.35

8

Some Final Thoughts

> *While some questions concerning the* luoshu *have been resolved, many mysteries still remain. In particular, three issues still linger. Why did the magic square gain such metaphysical prominence in China rather than the West? What happened to the* luoshu *as a visible symbol of harmony and cosmic balance? What is the ultimate significance of the* luoshu?

So far, we have traced the path of the *luoshu* from the banks of the Luo river, through the chambers of the *Mingtang* ritual hall, to the shuffling dance steps of Daoist priests and the star configuration of the Great Bear. The *luoshu*'s association with the theories of *yinyang* and *wuxing* emerged and its perceived status as a harbinger of human destiny became clearer. Later, Islamic cultures adopted the magic square and incorporated it into broader astrological theories. Before long, it was firmly enthroned in the Islamic world as an amulet and talisman which could affect the fortunes of its patrons. It is with this reputation that the *luoshu* arrived in Europe in approximately the twelfth century.

The *luoshu* has captured and held the imagination of many peoples in many forms and contexts. I hope that you have arrived at this point with a better appreciation of the *luoshu*'s mathematical significance, its relevance as a focus of Chinese cosmological and metaphysical theories, its capacity as a repository for occult beliefs, and its power as a symbol of cultural expression.

Why Did the Magic Square Originate and Flourish in China Rather Than in the West?

With strong Western traditions of number mysticism established by the Pythagoreans and their followers, why is it that the magic square first appeared in China, where mathematics was deemed a minor activity unworthy of scholarly pursuit?

Attempts to assert early European origins and conceptions of the magic square—for example, as a product of neo-Pythagorean experimentation by such individuals as Apollonius of Tyana, Theodorus of Asine, or Theon of Smyrna—falter under historical scrutiny.[1] One of the most interesting claims for European knowledge of the magic square focuses on the "SATOR square" of early Christian Rome.[2] Found scrawled on walls as a form of graffiti, the square is composed of letters which, in themselves, form five Latin words: *sator, arepo, tenet, opera,* and *rotas*.[3] The square is a two-dimensional palindrome: whether read from left to right, top to bottom, or from right to left, bottom to top, the same message is conveyed. See figure 8.1.

S	A	T	O	R
A	R	E	P	O
T	E	N	E	T
O	P	E	R	A
R	O	T	A	S

FIGURE 8.1

However, unlike a numerical square, the diagonals do not convey a meaning. Several messages have been assigned to this square. Some scholars have maintained that it conveyed statements of encouragement to a persecuted Christian community. The central configuration formed by the world *"tenet"* [to hold fast], forms a cross and if the letters comprising the square are rearranged, a *"Pater Noster"* [Our Father] cruciform anagram appears whose bounding letters A α, *alpha*, and O Ω, *omega*, proclaim the reign of Christ.[4] See figure 8.2.

Later, during the European Middle Ages, this same square of letters was considered a charm for inducing dancing among its viewers, detecting witches, and preventing disease.[5] Although the SATOR square is an intriguing device, it is not a magic square.

```
S   A   T   O   R              A
A   R   E   P   O              P
T   E   N   E   T              A
O   P   E   R   A              T
R   O   T   A   S              E
                                R
              A | P A T E R N O S T E R | O
                                O
                                S
                                T
                                E
                                R
                                O
```

FIGURE 8.2

As we have seen in chapter 6, the similarities between Pythagorean and early Chinese beliefs are striking.[6] Both peoples engaged in a philosophic search for the prerequisites of cosmic harmony and the importance of music as a conveyor of harmony. For the Pythagoreans, "all was number" and they rigidly structured their philosophical and metaphysical systems around this concept. Chinese metaphysical and philosophical theories are marked by a flexibility, a fluidity, whereby one concept flows into another.

The *luoshu* did not come into being *in toto*; rather, it evolved in several stages. First, it appears, that at a very early date, the geometric square as a symbol of the earth was incorporated into ritual structures and ceremonies. During the Warring States period, a tendency to divide special objects and categories into nine parts prevailed in philosophical thinking. The Earth-square was then divided into nine equal parts. This construction supplied a grid of nine cells, of which one occupied the strategic central position. Thus, the nine-celled square now became a symbol of a corporate unity dominated by a central presence. Its geometric symmetry served as a model for political and societal associations. When the magic square of order three was discovered, it was readily adopted into the awaiting grid of cells. Now geometric harmony was reinforced and strengthened by a numerical harmony emphasizing the importance of the numbers nine and five, which were already held in special reverence by the Chinese. The cycles of *yinyang* and *wuxing* were envisioned in the number patterns of the *luoshu*. The *luoshu* became a dynamic device, a *mandala*, which receptive viewers associated with various mystical theories. For early Chinese scholars, and eventually Daoist priests, the *luoshu* became a cosmogram capturing and projecting a world view based on balance and harmony. Thus, the *luoshu* evolved within Chinese society both as a concept and as a device and served as a repository for metaphysical beliefs.

In the Pythagorean world, there was no single concept as metaphysically comprehensive as the *luoshu*. The Pythagoreans held to the theory of crystalline spheres as a model for the universe; they associated five polyhedrons (the Platonic Solids) with the elements and the universe and employed the *tetractys* as their cosmogram. In the Pythagorean system of beliefs, the role of the *tetractys* most closely parallels that of the *luoshu*. But the *tetractys* was limited in its scope; it did not evolve or adapt to new theories. When Agrippa of Nettesheim presented his compendium of occult beliefs to a European audience in the sixteenth century, his depiction of the *tetractys* differed little from the one given by the Pythagoreans two thousand years earlier. Perhaps it was a reverence for the Master Pythagoras that caused the *tetractys*, as his personal creation, to be fixed in time. Thus, the *luoshu* presents a dynamic representation of metaphysical processes, quite different from Pythagorean depictions of cosmic harmony.

This different, dynamic way of looking at things was probably a factor in the development of the numerical magic square of order three. Cammann and Needham postulate that it was the early Chinese ability to express any number using only nine symbols that led them to a theory of magic squares.[7] This would certainly be a factor in perceiving and manipulating numerical patterns, but another aspect of early Chinese computation may also have contributed to the development of magic squares. Mathematical computation in ancient China was performed with a set of "counting rods." Users arranged these rods on a flat surface and moved them horizontally and vertically within the bounds of a rectangular or square matrix of available rod spaces.[8] This method of working within a square may have been a contributing factor in the Chinese discovery of magic squares.

What Happened to the *Luoshu* as a Visible Symbol of Harmony?

The *luoshu* became an adaptable cosmological device manipulated to conform to popular metaphysical and philosophical theories. Its number sequence, traced out, gave rise to the *yubu* ceremonies of Daoist ritual and when this path of progression to celestial and earthly harmony was graphically depicted, it became both a symbol associated with the *luoshu* and a talisman of good fortune itself. See figure 8.3. [9]

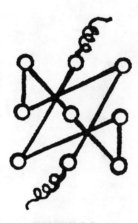

FIGURE 8.3

As previously discussed, the *hetu* diagram evolved from the *luoshu*. It was felt that the *hetu* presented a more cogent scheme for demonstrating the interactions of *wuxing* theory. In its numerical span the *hetu* differs from the *luoshu* in that it includes the number ten. Both the numbers five and ten represent Earth. In a *yinyang* context, the number ten is superfluous; numbers evolve or move towards their *yinyang* complement—the number added to which makes ten, so, 1 goes to 9, 3 goes to 7, and so on. Now, if this policy is followed and the *yin* numbers and the *yang* numbers within the *hetu* are partitioned from each other, a spiral configuration of lines is formed and the circle containing the *hetu* is divided into two complementary regions. If these regions are given opposite colorings, say white and black, the figure that emerges is the *taijitu*, the traditional, recognized symbol of *yinyang* interaction. This symbolic evolution is illustrated in figure 8.4.

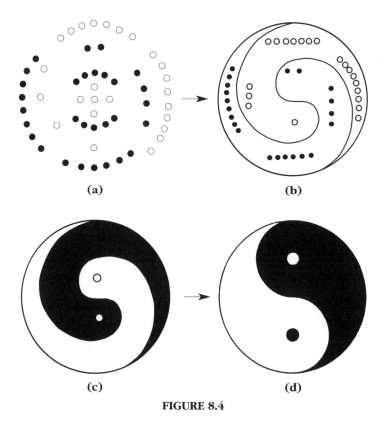

FIGURE 8.4

Scholars can trace the emergence of the *taijitu* as a motif and symbol of cosmological harmony to about the tenth century. Thus, it seems highly likely that the *luoshu* finally evolved into the visually more appealing circular symbol of change and harmony, the *taijitu*, which remains with us today.

What is the Ultimate Significance of the *Luoshu?*

So what is the lasting significance of the *luoshu*? Certainly it has served many different functions. James Legge, one of the first Western scholars to delve into Chinese classical lore, dismissed it as an "arithmetical puzzle" and the "reductio ad absurdum of Lo writing."[10] Marcel Granet, an anthropologist with a more sympathetic view of Chinese mysticism, urges a deeper study of the beliefs and practices surrounding the number square as a means of understanding early Chinese society.[11] Outside of its Chinese context, the magic square has still remained an object of controversy and fascination. Hermann Schubert, writing on mathematical recreations in the early twentieth century, noted that magic squares have "led many a man to believe in mysticism." He advised that there was "nothing magical about them" and then went on to describe them as examples of a "symphonic harmony in mathematics."[12] Martin Gardner, the current "Dean of Mathematical Problem Solvers," considers the *luoshu* "one of the most elegant patterns in the history of combinatorial number theory."[13] While these contrasting opinions are diverse and based on different perspectives and realms of experience, they all share one feature: they reflect Western expectations. The true value of the *luoshu* can only be appreciated in the context of ancient China. How did it serve Chinese society?

In the preceding discussions of the uses of the *luoshu*, two words frequently surfaced: "cosmogram" and "mandala." A cosmogram is a map or diagram of the "universe"—an illustration of a particular peoples' world view. A mandala is a ritual geometrical diagram with symbolic significance for the viewer. Some examples of mandalas are the Hindu meditation aid referred to as the *sriyantra*, the Buddhist "Wheel of the Universe," and the Aztec "Great Calendar Stone."[14] A very powerful and influential cosmic mandala for the people of Mesoamerica was a cross-like configuration that divided the world into five regions, one for each major direction. See figure 8.5. Each direction had a name, a color, and other attributes associated with it.[15] The central region, the "navel of the Earth," was believed to be the domain of the fire deity.

For the Chinese, the *luoshu* served in the capacity of both a cosmogram and a mandala.

As a map of the universe, its properties were relevant to the time of its conception. Heaven and Earth were different: Earth, represented by a square, was finite and rigidly bounded by four principal directions; Heaven circumscribed Earth as a circle and, as such, was not bounded by directions but extended in all directions. Both Heaven and Earth revolved around a great central axis. Motion and cyclic change were integral parts of the Chinese world picture. The universe was a vast organic entity in a state of constant flux. *Yinyang* and *wuxing* theories were formulated to explain the cycles of change and made consistent with the workings of the *luoshu*. Nor was the Earth, as envisioned by Chinese cosmic thinkers, a uniform entity. It possessed features— provinces, mountains, rivers, and so on—that affected the lives of the Chinese people. The *luoshu* with its nine cells accommodated these fea-

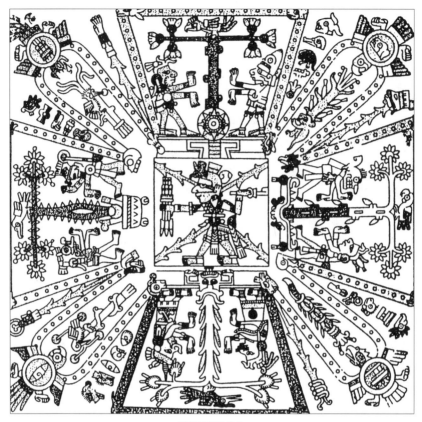

FIGURE 8.5

tures, primarily by expressing concepts of centrality and control. Thus, within the realm of the *luoshu*, the central cell—representing the Chinese people, China, the Middle Kingdom, or the Emperor—served as a unifying link with the peripheral regions that represented lesser beings or institutions. A hierarchy of control and political domination was inherent in the *luoshu* structure. The tendency toward psychocentrality was not unique to the Chinese; an Earth-centered universe held the attention of Western philosophers and theologians for two millennia. In the *Topographia Christiana* of Cosmos of Alexandria (ca. +548), the sacred city of Jerusalem dominated the center of a rectangular Earth. Interestingly, the phrase "the four corners of the Earth" is still a common expression in English.

Early scientific expression often took the form of interpretive number statements. Theories were modeled by the interaction of numbers and their properties. The *luoshu* display of the numbers 1 through 9 is such an articulation. Symmetry of sums, *yinyang* balance, the paths of the *wuxing* cycles, and the constancy of the magic sum were all numerically evident and relevant to the "believing" observer. In ancient societies, numbers provided an empowering means to reach beyond human experience, as noted by Granet:

> It is by means of numbers that one finds a suitable way to represent the logical sectors and concrete categories that make up the universe. . . . In choosing for them one or another disposition which permit them to demonstrate their interplay, one believes he has succeeded in rendering the universe at once intelligible and manageable.[16]

While to the modern, scientifically trained mind, the theories of *yinyang* and *wuxing* may seem like so much metaphysical claptrap, they must be viewed from the perspective of their times. The dualistic binary encoding of the *yinyang* and the *bagua* differ little in their conception and use from the processes of a modern electronic computer where information is transformed into binary messages, stored, and eventually acted upon. *Wuxing* theory with its inclusive categories and system of correlational reckonings foreshadowed the popular European doctrine of Four Elements and Humours by a thousand years. Examples of Four Elements correlations are given in table 8.1.

Well into the seventeenth century, Four Elements theories clouded European perceptions of terrestrial phenomena and human behavior. Johann Kepler (1571–1630), in his quest for a more rational system of planetary motion, was plagued by the legacy of the Pythagorean crystal sphere theory. For many years, and with great frustration, he attempted

TABLE 8.1

Element	Physical State	Season	Color	Direction	Humour	Human Trait
air	gas	Spring	yellow	east	sanguine	intellect
fire	agent of change	Summer	orange	south	choleric	anger
earth	solid	Autumn	green	north	melancholic	patience
water	liquid	Winter	blue	west	Phlegmatic	sensitivity

to fix his observational findings to a system of celestial circles.[17] Eventually, he abandoned this effort and began to use mathematics to calculate elliptical orbits for the planets. Thus, even in fairly recent European experience, mystical thinking and scientific outcomes were frequently intertwined.

The Chinese theories that were embodied in the *luoshu* were based on observations of nature, the formation of hypotheses, the establishment of premises, and a system of deductive reasoning. Indeed for Chinese scholars of the early Han period, chains of inductive reasoning flowed easily from *yinyang* and *wuxing* speculations. Causal relationships were limited to preordained cycles of change. What keeps these theories from being recognized as "scientific" in the Western sense is a lack of experimentation and an intrusion by the observers into the processes being observed. Modern science, boldly challenging the fates that shape human existence, only became commonplace in Europe after the establishment of the "Scientific Method" attributed to Galileo Galilei (1564–1643). In light of this perspective (bias?), *yinyang* and *wuxing* are deemed proto-scientific theories and the *luoshu* as a symbolic and operative repository for these theories, a proto-scientific device.

In a more human and less analytical sense, the *luoshu* served a broader purpose in early Chinese society as a visual statement of cosmic and personal harmony, a symbol of order. The shape, the configuration of numbers, and their position relative to each other served as a way to contemplate our place in the universe. This function of the *luoshu* was often embedded in art motifs, the design of game boards, cosmic mirrors, and fortune-telling instruments. As a stimulus for philosophical and metaphysical speculation, the *luoshu* also served as a mandala.

But perhaps its most dominant characteristic was its capacity to absorb and symbolize the major cosmological theories of early China. While these theories were apparently appended to the square over a period of several hundred years, they found accord within the awaiting matrix. Whether this synthesis was by design or coincidence may never be known; however, the status of the *luoshu* as a symbol of Chinese cosmological thought cannot be denied. This feature, along with its historical preeminence as the first magic square as well as the philosophical and psychological reverence accorded to it by so many peoples of the world, make the *luoshu* a truly magical entity, the most magic of all magic squares.[18]

Epilogue

Deep Background: How Magic Squares Fit into the Big Picture

Since earliest times, we human beings have sought to find consistency in our environment. Whether it was learning to recognize the paths taken by the wandering sky-god planets[1] or discovering the predictability of the seasons or the process of night following day, the "sameness," regularity, and knowledge of what could be expected provided comfort to our human ancestors in an otherwise uncertain and hostile world. A calendar could then control and dictate activities—some degree of order was established. To a large extent, existence and survival became contingent upon such predictions and orientations. Within social groups, individuals emerged who recognize the unfolding patterns around them, partially understood their consequences, and advised their society accordingly.[2] The fact that some early world-observers were in a position to acquire knowledge about future events and the skills to use certain tools and measuring devices was one factor that contributed to their emergence as a respected priestly or ruling class.

Ancient priests or shamans surrounded themselves with institutions, rituals, and artifacts that directly related to the natural phenomena they deemed important. In particular, cosmic events—the movement of the moon and observable planets, the appearance of a comet—were carefully scrutinized, interpreted within the context of existing cosmological beliefs, and acted upon. The tools used for observing and measuring natural phenomena became part of the ritual paraphernalia and acquired a mystical aura of their own. For example, ancient observers discovered that they could use a staff placed vertically in the ground to gauge the changing length of a shadow caused by the sun and thus also the changing position of the sun relative to the Earth. They then figured out that they could use the same staff (or "gnomon") to determine the time of the winter solstice. Eventually people came to treat the staff itself as a sacred instrument for insuring the lengthening of the days.[3] Similarly, a configuration of rocks that permitted the monitoring of celestial alignments might have become an aid for keeping track of the heavenly gods who traveled the night sky. Primordial monuments that served such functions include Stonehenge, on the plains of Salisbury in England, and the ingenious Sun Dagger construction of the Anasazi people, former cliff dwellers of the Chaco Canyon of the Southwestern United States.[4] An examination of other such devices and monuments from the ancient world can provide valuable insights into the prescientific thinking,[5] beliefs, and methodologies of our ancestors.

Within this great quest to understand and control the world around them, early peoples used certain concepts that reflected desired expectations and behavior. Principal among these concepts was that of harmony—individuals and society as a whole needed to be in harmony with the cosmos and the cosmos needed to be in harmony with them. People could experience harmony in several ways: (a) through a feeling of inner peace, a state of easiness marked by a lack of anxiety, or (b) through visual or aural perceptions that convey symmetry and balance and contribute to a sense of beauty. The ancient Greeks associated harmony with the "music of the spheres," the imagined whirling of the planets (moon and sun included) in their circular orbits about the Earth.[6] Most traditional peoples of recent times live lives dependent on harmonious coexistence with nature. Often their concepts of harmony are reflected in music, dance, and visual art forms.

Certain preferred geometric shapes and designs have emerged that reflect a group's concept of harmony. For example, the Osage people, native to the plains of North America, set up their camps in a circle. For them, the camp-circle was also a reflection of the cosmos at large: the half of the tribe residing in the northern side of the circle represented Heaven while the half living in the southern section symbolized Earth. Heaven and Earth were unified within the circle of the camp. It is easy to understand why the circle might be associated with the concept of harmony. A circle has no beginning or end; it repeats itself continually in a perfect cycle—always changing but always the same; the circle is evenly spaced around its center, providing a paramount example of balance. Philosophers and mystics in many cultures believed that the circle held psychological and metaphysical significance.[7] But in still other cultures, the geometric shape that best epitomized harmony was a square encompassing the "four corners of the Earth," an equilateral triangle depicting the Christian Trinity, or a three-dimensional figure such as a sphere.[8]

In addition to supplying geometric forms that can be symbolically associated with harmony, mathematics can assist people in organizing the world into an understandable system in other ways. For instance, early peoples used number to order their environment. Number, when recognized as a property of an object or objects, conveys a certain permanence. For we who live on Earth, there is one moon and one sun—we see them today and would expect to see each one tomorrow. Because we can count limbs and digits, we have the expectation that every person we see will have two arms and five fingers on each hand, barring injury. Quantification enabled people to sort objects into real or imagined categories and relate the categories to one another in a linear or hierarchical relationship.

Numbers, as powerful descriptors, often took on a mystical significance and ritual meaning.[9] Certainly the quantifier "one" as a conveyor of uniqueness distinguished an individual object, whether that individual was a person, an animal, or an inanimate object. The number one, as the first counting number, occupies a certain position of importance—the other numbers in the system depend on it for their existence; thus, in many societies, the number one takes on a certain ritual meaning, one of importance or leadership.[10] Similarly, the number two represents an "otherness" which can be associated with opposition, a challenge, or an inadequacy—in some modern Western societies, "being number two" indicates being inferior. Thus, numbers take on meanings that extend beyond their enumerative capacity.

The concept of "lucky" and "unlucky" numbers remains with us today, although its origins can be traced back thousands of years.[11] Much of the number lore evident in Western civilization evolved from the Pythagorean traditions of ancient Greece. The cult followers of the philosopher Pythagoras of Samos (ca. 585–ca. 500 B.C.E.) proclaimed "all is number" and developed a complex numerology centered on human existence and activity. One Pythagorean set of number relationships is as follows: two is the female number, three is the male number; the sum or "coming together" of two and three is five, thus five is the number for marriage, and the product of two and three is six which became the number of procreation.

Similarly, the English expression "square deal" refers to a transaction based on fairness, an equal distribution of the items or services in question. This phrase evolved from the Pythagorean association of the number four with the concept of justice. Because four is the first number composed of equals coming together, namely, 2×2, we would describe it as the first square number. Vestiges of ancient number theories and beliefs are omnipresent even today.

In modern times, the search for harmony has resulted in an exciting new field of mathematical endeavor, chaos theory. Mathematicians working in this branch of applied mathematics are attempting to unravel the mystery of chaos (chaos being a state of disorder or uncertainty) with the aid of computer simulations and astute mathematical modeling techniques.

Following the *luoshu*'s path has provided a revealing glimpse of the human capacity to contrive meaningful, far-reaching, yet often flawed "scientific theories" within a limited realm of comprehension where facts, beliefs, and superstitions all become intertwined. Ultimately the *luoshu,* in its geometrical and numerical conception, marks an effort in the ongoing search for harmony.

Notes

Prologue

1. Simon de la Loubère, *A New Historical Relation of the Kingdom of Siam by Monsieur de la Loubère, Envoy-Extraordinary from the French King to the King of Siam in the years 1687 and 1688*, trans. A. P. Gent (London: Horne Saunders & Bennet, 1693; from the French edition, Paris, 1691; reprinted John Villieis, ed., Bangkok: White Lotus, 1986).

2. The book's reference number was D899.64C148. It was an English edition of 1693. An inquiry in recent years indicates the book has disappeared from the library's holdings.

3. This study was published: Frank Swetz, *Mathematics Education in China: Its Growth and Development* (Cambridge, MA: MIT Press, 1974).

4. Schuyler Cammann: "The Evolution of Magic Squares in China," *American Oriental Society Journal* 80 (1960): 116–24; "The Magic Square of Three in Old Chinese Philosophy and Religion," *History of Religions* 1 (1961): 37–80; "Old Chinese Magic Squares," *Sinologica* 7 (1962): 14–53; "Islamic and Indian Magic Squares, Part I," *History of Religions* (Feb. 1969): 181–209; "Islamic and Indian Magic Squares, Part II" (May 1969): 271–99.

5. Marcel Granet, *La Pensée chinoise* (Paris: La Renaissance du Livre, 1934).

6. See note 4 in the prologue for a list of magic squares articles by Schuyler Cammann.

7. Ho Peng Yoke, "Magic Squares in East and West," *Papers on Far Eastern History* 8 (1973): 115–41.

8. Lars Berglund, *The Secret of Luo Shu: Numerology in Chinese Art and Architecture* (Lund: Lund University, 1990).

Chapter 1: The Journey Begins

1. Noël Golvers, *The Astronomia Europaea of Ferdinand Verbiest S. J.* (Dillingen, 1687), Monumenta Serica Monograph Series (Nettelal: Steyler Verlag, 1993).

2. In general, Chinese names and terms will be transliterated using the Pinyin system of romanization. For much of the twentieth century the Wade-Giles system was employed for rendering Chinese into English. In referring to proper names and titles and certain concepts that have become well known by their Wade-Giles spellings the older system will be used.

3. See Lillian Too, *Feng Shui* (Kuala Lumpur: Konsep Books, 1993).

4. The quote is from Black Elk, a Sioux chief, lamenting in the nineteenth century on the prospects of reservation life:

> You have noticed that everything an Indian does is in a circle, and that is because the Power of the World always works in a circle, and everything tries to be round. . . . But the Waischus [white men] have put us in square boxes. Our power is gone and we are dying, for the power is not with us anymore.

See John Neihardt, *Black Elk Speaks* (Lincoln, NE: University of Nebraska Press, 1972), 198–99.

5. Dr. Golvers is attached to the Ferdinand Verbiest Foundation, Leuven.

6. Noël Golvers also referred me to Plate 213 in the contribution by Howard L. Goodman, "Paper Obelisks: East Asia in the Vatican Vaults," *Rome Reborn: The Vatican Library and Renaissance Culture*, ed. Anthony Grafton (Washington, DC: Library of Congress, 1993), 251–93.

7. Personal inspection of the papers convinced Dr. Golvers that they were not written by Verbiest.

8. Granet also theorized upon this relationship. See *La Pensée Chinoise* (Paris: Renaissance du livre, 1934).

9. The Jesuit incursion into China is a fascinating episode in the history of cultural interaction. Some of this story is told in: David Mungello, *Curious Land: Jesuit Accommodation and the Origins of Sinology* (Stuttgart: Franz Steiner Verlag, 1985) and Jonathan Spence, *The Memory Palace of Matteo Ricci* (New York: Viking Penguin, 1984).

10. The growth of Leibniz's interest in China is best described in Gottfried Wilhelm Leibniz, *Writings on China*, trans. Daniel J. Cook and Henry Rosemont (Chicago: Open Court, 1994).

11. See Golvers, *The Astronomia Europaea of Ferdinand Verbiest*, for a discussion of Verbiest's work in China.

12. Claudia von Collani, *P. Joachim Bouvet S. J. Sein Leben und sein Werk*, Monumenta Serica Monograph Series x vii (Nettetal: Steyler Verlag, 1985)

13. Leibniz discussed his binary system with correspondents as early as 1679 but formally made his announcement to the scientific community at large

in 1703 in "Explication de l'arithmétique binaire," *Memoires de l'Académie Royale* (1703), 85–89. See Anton Glaser, *History of Binary and Other Nondecimal Numeration* (Southampton, PA: Tomash, 1981).

14. Leibniz, *Writings on China,* 73.

15. See David Mungello, *Leibniz and Confucianism: The Search for Accord* (Honolulu: University of Hawaii Press, 1977).

Chapter 2: The Chinese Origins of the *Luoshu*

1. The "Three Cultural Heroes" are Fuxi, also considered to be the patron of animals; Shen Nung, inventor of agriculture, commerce and herbal medicine; and Huang Di, the Yellow Emperor, inventor of writing and weapons. These heroes were believed to flourish in the third millennium B.C.E. See discussion of Fuxi, Shen, and Yao in *Sources of Chinese Tradition*, vol. 1, ed. William de Bary (New York: Columbia University Press, 1970), 197–98.

2. Yao the Virtuous Emperor abdicated in favor of the wise commoner Shun; Yu the Great was the tamer of floods. Archaeologists have identified the Xia "Dynasty" with a late stage of neolithic Longshan culture.

3. *Yijing* has no specific author. Like many ancient Chinese classics it organically grew out of change itself, as a collective tradition of many anonymous shaman. The tradition author of the *Yijing* is Fuxi or Confucius. The book contains the basic theory of *bagua* and the sixty-four hexagrams and the *yin-yang* deviation scheme.

4. Water, its functions and uses, figure prominently in early Chinese myths and beliefs. For a discussion of water associations, see Evelyn Lip, *Chinese Geomancy* (Singapore: Time Books International, 1979), 10.

5. Winds blow from Russia across the dusty plateaus of central north-west China carrying off the yellow loess topsoil. The Yellow River and the Yellow Sea owe their color to this dust. In turn, to Yellow River gives life to the agricultural lands of north China. Yellow was the traditional color associated with ancient China.

6. See Karl A. Wittfogel, "Die Theorie der orientalischen Gesellschaft," *Zeitschrift fur Sozialforschung* 7 (1938): 90, and *Oriental Despotism* (New Haven: Yale University Press, 1957). While many contemporary anthropologists do not favor Wittfogel's theories, I find them particularly suited for Chinese society. To better understand the importance of the image of water in ancient Chinese thought, see Sarah Allan, *The Way of Water and Sprouts of Virtue* (Albany: SUNY Press, 1997).

7. There were three major schools of cosmological thought in ancient China: The *gaitian* (hemispherical dome) school; the *huntian* (celestial sphere) school and the *xuanye* (infinite empty space) school. The *gaitian* theory holds that the heavens are like an open umbrella over an inverted bowl-like earth—

that is the analogy of the tortoise shell. *Huntian* followers believe the heavens are a sphere, like an egg, and the earth sits within this sphere like a yoke. *Xuanye* belivers consider the earth a "gram of rice" in an infinite empty space. See discussion Ho Peng Yoke, *Li, Qi and Shu: An Introduction to Science and Civilization in China* (Hong Kong: Hong Kong University Press, 1985), 126–30.

8. See S. Cammann, "The TLV Pattern on Cosmic Mirrors of the Han Dynasty," *Journal of the American Oriental Society* 68 (1948): 159–67, and Sarah Allan, *The Shape of the Turtle: Myth, Art and Cosmos in Early China* (Albany: SUNY Press, 1991).

9. Qin Shihuangdi, emperor from 221 to 210 B.C.E. unified China, started work on the Great Wall and initiated social and political reforms. A despotic ruler, Qin ordered a "burning of the books" to eliminate possible threatening political theories from becoming popular.

10. Confucianism and Daoism became the prevailing ethical/religious doctrines of Chinese society. Mohism, based on the teachings of the philsopher Mo Tzu (470-391 B.C.E.), eventually lost its influence. Buddhism was imported into China in about the first century and has retained a spiritual influence.

11. One of the great quests of Daoist alchemists was the seeking of an elixir of life.

12. The founders of Daoism are usually considered to be Laozi and Zhuang Zi. Zhuang is known to be an historical personage; however, Laozi's actual existence is debatable.

13. See discussion of this issue in Schuyler Cammann, "The Magic Square of Three in Old Chinese Philosophy and Religion," *History of Religions* 1 (1961): 37–80, 44.

14. William E. Soothill gives a fascinating study of the institutions of the *Mingtang* in *The Hall of Light: A Study of Early Chinese Kingship* (New York: Philosophical Library, 1952).

15. This description was used by many later commentators on the "Nine Halls" diagram. Its analogy is better understood when methods for construction the Nine Halls/*luoshu* diagram are considered. See note below.

16. Zheng's representation of the *luoshu's* numbers by knots in a cord was intended to associate the configuration with more ancient times, a common literary device used by commentators on old works to increase their apparent age and thus importance in the eyes of a Chinese reader. The *Yijing* notes that "in Early Antiquity, knotted cords were used to govern with. . . ." For a discussion on the use of these knotted cords in China, see Jean-Claude Martzloff, *A History of Chinese Mathematics* (New York: Springer-Verlag Berlin Heidelberg, 1997), 180. A similar type of knotted cord for keeping track of computations, called a *quipu*, was used by the Incas of Peru. For more on the subject, see Lind Mae Diana, "The Peruvian Quipu," *Mathematics Teacher* 60 (Oct. 1967): 623–28; reprinted in *From Five Fingers to Infinity*, ed. Frank Swetz (Chicago: Open Court, 1994), 80–85.

17. Ho Peng Yoke, "Chinese Magic Squares: Mathematics, Myth and Philosophy," *Kertas-kertas Persidangan Antarabangsa Pengajian Tionghoa* [*Collected Papers of the International Conference on Chinese Studies*] (Kuala Lumpur: November 20–21, 1993), 346–72.

18. The complete work is translated and commented upon by Lam Lay Yong, *A Critical Study of the Yang Hui Suan Fa: A Thirteenth-Century Chinese Mathematical Treatise* (Singapore: Singapore University Press, 1977).

Chapter 3: *Yinyang, Wuxing,* and Key Numbers

1. For example, in Singapore and Hong Kong, people do not want the number 4 to appear in the number on their cars' license plate numbers. Popular Chinese numerology and its beliefs are discussed in Evelyn Lip, *Chinese Numbers: Significance, Symbolism and Traditions* (Singapore: Times Books International, 1992).

2. This phenomena is discussed by Derk Bodde, "Types of Chinese Categorical Thinking," *Journal of the American Oriental Society* 59 (1939): 201–21.

3. This theory is more fully discussed and developed in Lam Lay Yong and Ang Tian Se, *Fleeting Footsteps: Tracing the Conception of Arithmetic and Algebra in Ancient China* (Singapore: World Scientific, 1992).

4. Zero, as a numerative symbol was not employed in China until the time of the Tang Dynasty (618–906). On counting board operations, an empty space was left to mark a state of "nothingness" for the decimal position.

5. In a decimal-based numeration system, the numerals 1 to 9 count the units, tens, hundreds, and so on up; thus 153 is 1 hundred, 5 tens, and 3 units. "Ten," "hundred," and "thousand" are names of the decimal groupings.

6. Passages from the *Yugong* chapter of the *Book of History*, translated in Ming-chong Hwang, "*Ming-tang*: Cosmology, Political Order and Monuments in Early China," unpublished Ph.D. dissertation (Department of East Asian Languages and Civilizations, Harvard University, 1996), 426.

7. For a fuller discussion of the psychology and mechanics of finger counting see Georges Ifrah, *From One to Zero: A Universal History of Numbers* (New York: Viking Penguin, 1985).

8. In representing a number, the numerals shown in Figure 3.1 are used to represent coefficients of 10^{2n-2} n = 1, 2, ... A variation of them:

is used for coefficients of 10^{2n-1} n = 1, 2, ... Thus in this system, 4716 would be represented as

9. See J. M. Pullan, *The History of the Abacus* (New York: Frederick A. Praeger), 1969.

10. Ho, *Li, Qi, and Shu*, 21.

11. See discussion of Zou Yan's work in Fung Yu-lan, *History of Chinese Philosophy*, vol. 1, trans. Derk Bodde (Princeton: Princeton University Press, 1955), 161–62.

12. Another association for "nineness" and mathematics is the title of the most influential of all Chinese mathematical classics, *Jiuzhang suanshu* [The Nine Chapters of the Mathematical Art] (ca. 100 B.C.E.) which summarizes all of mathematics under nine topics (types of applied mathematics). For more information on early mathematics teaching and learning in China, see Frank Swetz, *Mathematics Education in China: Its Growth and Development* (Cambridge, MA: MIT Press,1974).

13. The Chinese name for China, *Zhongguo*, means "Middle Kingdom" or "Central States" throughout history they have considered it, at least psychologically, at the center of the universe—between Heaven and the "barbarians." In their conception of the Earth as a great square, they envisioned it surrounded by "Four Seas" which were not oceans but rather regions populated by uncivilized barbarians. The East contained nine kinds of *Yi* barbarian; the West, seven kinds of *Rong* barbarian; the North eight kinds of *Di* barbarian and the South six kinds of *Man* barbarian. When the Chinese first came in contact with Western nations, such an attitude, one of ethnocentric superiority did not win them friends.

14. For a discussion of various map orientations see B. L. Gordon, "Sacred Directions, Orientation, and the Top of the Map," *History of Religions* 10 (1971): 211–27. While the Chinese utilized directional orientations based on five principal directions, other traditional people employed even more directions. For example, the Hopi people of Southwest North America had six major directions: four horizontal directions and the zenith and the nadir. Discussions of traditional classification systems are given in Claude Lévi-Strauss, *The Savage Mind* (Chicago: University of Chicago Press, 1966).

15. See discussion by Weihang Chen, *"Yinyang,"* *Encyclopedia of the History of Science, Technology and Medicine in Non-Western Cultures*, ed. Helaine Selin (Boston: Kluwer Academic Publishers), 1045–46.

16. Translated in W. Theodore de Bary, Wing-tsit Chan, and Burton Watson, *Sources of Chinese Tradition*, vol. 1 (New York: Columbia University Press, 1970), 192–93.

17. Passages in the *Daodejing* [Writings of Lao Zi] provides a more extensive list of contrasts:

Yang	**Yin**
1. Heaven	Earth
2. Spring	Autumn

3.	Summer	Winter
4.	Day	Night
5.	Big states	Small states
6.	Important states	Insignificant states
7.	Action	Inaction
8.	Stretching	Contracting
9.	Ruler	Minister
10.	Above	Below
11.	Man	Woman
12.	Father	Child
13.	Elder brother	Younger brother
14.	Older	Younger
15.	Noble	Base
16.	Getting on in the world	Being stuck in one's position
17.	Taking a wife, begetting a child	Having a funeral
18.	Controlling others	Being controlled by others
19.	Guest	Host
20.	Soldiers	Laborers
21.	Speech	Silence
22.	Giving	Receiving

18. As quoted from a passage from the lost book *Classic of the Nine Halls*. See Schuyler Cammann, "The Magic Square of Three in Old Chinese Philosophy and Religion," *History of Religions* 1 (1961): 37–80.

19. See Ang Tian Se, "Five Phases (*Wuxing*)," *Encyclopaedia of the History of Science, Technology, and Medicine in Non-Western Cultures*, ed. Helaine Selin (Boston: Kluwer Academic Publishers), 332–33.

20. The popular theory held in Europe up through the time of the Renaissance was that all matter was comprised of four elements: earth, air, fire, and water. This theory is attributed to Empedocles of Acragas (ca. 490–435 B.C.E.), a Pythagorean.

21. Traditional Chinese music was pentatonic in nature. Its five tones were said to represent the emperor, the minister, the people, the affairs of state and material objects. Using the Western musical stave, equivalent notes would be C, D, E, G, A.

22. A modern observer of Chinese life may wonder why rice is not included in this list.

It must be remembered that rice is a principal grain in southern China and that early Chinese civilization developed mainly in northern China where millet and wheat still remain the most popular grains.

23. *Zou's* theory noted that:

Each of the Five Virtues [*wuxing*] is followed by one it cannot conquer.
The dynasty of Shun was ruled by the virtue of Earth, the Xia dynasty by

the virtue of Wood, the Shang dynasty by the virtue of Metal and the Zhou dynasty by the virtue of Fire.

Further in the text, he indicates that the rule of Zhou will be terminated by Heaven acting through water. Thus the cycle for the decline of dynasties is: water–fire–metal–wood–earth.

24. Ho, *Li, Qi and Shu*, 20.

25. For information on the swastika in Chinese thought see P. J. Loewenstein, "Swastika and Yin-Yang," China Society Occasional Papers (London: China Society, 1942). The symbol's use extends back to ancient times in China and can be traced to Sumeria (3000 B.C.E.). In the Chinese context, the symbol served as a general superlative with a spectrum of meaning centered around power, energy, and migration—all of which find appropriate interpretations in the *luoshu* context. For further information on the swastika motif in early Chinese art and architecture, see Lars Berglund, *The Secret of the Luo Shu*, 281–82. The swastika as a special symbol in other cultures is discussed in J.C. Cooper, *An Illustrated Encyclopaedia of Traditional Symbols* (London: Thomas and Hudson, 1978), 164–66.

26. Passages from *Chunqiu fanlu* (ca. 135 B.C.E.) as given in Ho, *Li, Qi and Shu*, 18.

27. For further discussion of the *hetu*, see Michael Saso, "What is the Ho-t'u?" *History of Religions* 17 (1978): 399–416.

Chapter 4: The *Luoshu* in Cosmic Ritual

1. *Yang* forces were believed to flow from south to north. For a comprehensive description to the temple complex see Nagel's *Encyclopedia-Guide: China* (Geneva: Nagel Publishers, 1984), 540–46.

2. After 1911, the emperor no longer performed ceremonies at the Temple of Heaven. On October 12, 1912, the Republican government declared Chinese National Day and opened the temple to the public and a minister performed a sacrifice to the "Supreme Lord" on behalf of the president. On the 1914 winter solstice, president Yuan Shi kai, who hoped to restore the empire, prayed at the temple.

3. Zou also theorized that the "Middle Kingdom" itself occupied one-ninth of the Red Continent, which was one-ninth the land mass of the world. Thus, the world could be divided into eighty-one subdivisions. See p. 23 above.

4. See, for example: Henri Maspero, "Le Ming-Thang et la Crise Religieuse Chinoise avant les Han," *Mélanges Chinois et Bouddhiques* 9 (1951): 1–71; William E. Soothill, *The Hall of Light: A Study of Early Chinese Kingship* (New

York: Philosophical Library, 1952); Hwang Ming-Chong, "Ming-tang: Cosmology, Political Order and Monuments in Early China," unpublished Ph.D. dissertation (Harvard University, Department of East Asian Languages and Civilization, 1996).

Cosmically significant elements can be found in the structures: temples, palaces, and monuments of many ancient societies. It is interesting to note that in the Vedic tradition of Hinduism, temple construction is based on a square, the symbol of the Earth. This square, known as *Vastu-Purusha-mandala,* is then subdivided into a series of smaller squares, which in turn lend themselves to the structural design of the temple. One of the basic schemes of subdivision is into a square of nine cells, of which the central one, being most sacred, is reserved for Brahma and represents the center of the world. The eight peripheral sub-squares designate the cardinal regions of the Earth. The similarity to the *luoshu* scheme is striking. For more information on ritual architecture see Titus Burckhardt, *Sacred Art in East and West: Principles and Methods* (Pates Manor, U.K.: Perennial Books Ltd., 1967).

5. See Hwang, "Ming-tang," 27. Excavations at Xi'an have revealed the foundations of the *Mingtang* built by the Han Emperor Wang Mang. See also Wang Zhongshu, *Han Civilization* (New Haven: Yale University Press, 1982), 39.

6. For example, *Dadai liji* [Record of Rites by Dai the Elder] (ca. 100 B.C.E.).

7. As given in a later commentary of *Dadai liji.* Translated in John B. Henderson, *The Development and Decline of Chinese Cosmology* (New York: Columbia University Press, 1984), 78.

8. A reconstruction of Wang Mang's *Mingtang* is given in Wang Shiren, *Kao Gu* (1963), figure 20.

9. As given in Soothill, *The Hall of Light,* 34. The square in figure 4.2 has been rotated 180 degrees from Soothill's illustration in order to conform to the traditional Chinese practice of placing the *south* at the top of a map.

10. As translated and given in Soothill, *The Hall of Light,* 30.

11. Ibid., 33.

12. Ibid., 37.

13. Ibid., 39.

14. Ibid., 40.

15. Chinese musical instruments are divided into eight classes depending on the material of their construction. These eight groups of instruments were correlated with the *bagua.* The instrument for the transition of spring to summer was the "wooden fish," a type of flute. Emperors possessed an imperial pitch pipe. It was a task of court musicians and astrologers, upon the emperor's request, to recalculate the proper length of this pipe to insure that the emperor would be in harmony with the universe. See the discussion of Chinese music in Rita Aero, *Things Chinese* (Garden City, NY: Doubleday & Company, 1980).

16. A detailed discussion of this ceremony is given in Hwang, "Mingtang," 403–73.

17. Ibid., 444.

18. As given in Hwang, "Mingtang," 47–48.

19. Heaven was divided into nine regions: the center, the Lathe Heaven; the Eastern, Blue Heaven; northeast, the Changing Heaven; north, the Black Heaven; northwest, the Dark Heaven; west, the Luminous Heaven; southwest, the Vermilion Heaven; south, the Burning Heaven; and southeast, the Yang Heaven. Further, both Heaven and Earth were thought to possess nine levels each. The bottom Earth-level was the land of the dead.

20. In 9 C.E., Emperor Wang attempted to organize a land-holding system based on the ancient "well-field" scheme of nine units. His reforms also included the nationalization of all land, abolition of private land-holding, and the prohibition of the sale of land or slaves. The attempts at reform failed. Emperor Wang built his own *Mingtang* in what is today Xi'an.

21. We recognize these stars as the Big Dipper. The old star manual *Xingjing* indicated *Beidou* originally consisted of nine stars—two of which faded over time. See discussion in Ho Peng Yoke, "Chinese Magic Squares: Mathematics, Myth and Philosophy," *Kertas-kertas Persidangan Antarabangsa Pengajian Tionghoa*, International Conference on Chinese Studies (Kuala Lumpur: November 20-21, 1993), 346–71.

22. Ho, "Chinese Magic Squares," 366. There exists a variety of Daoist ritual movements/dances, *bugang*, "walking the guideline," of which the *yubu* comprises one set. In turn, there are several forms of *yubu*. In the discussion, we are concerned with the *yubu* movements associated with the *luoshu*. For detailed information on the subject, see Poul Andersen, "The Practice of Bugang," *Cahiers d'Extrême-Asie* 5 (1989-90): 15–53.

23. The procedure for choosing rods is described in detail in "I-Ching," *Encyclopedia of the Unexplained: Magic, Occultism and Parapsychology*, ed. Richard Cavendish (New York: McGraw Hill, 1974), 122–25.

24. The hexagrams are tabulated and interpreted in Ho, *Li Qi and Shu*, 36–41. Gottfried Wilhelm Leibniz (1646–1716) learned of the existence of the hexagrams from Jesuit missionaries in China. He had developed a binary arithmetic and noted the resemblance between the hexagrams and his binary notation. Leibniz attached a religious significance to his findings, pointing out, "All combinations arise from unity and nothing, which is like saying that God created everything from nothing." Believing that this theory, so like that accepted by the Chinese, would help convert the Emperor to Christianity, he conveyed it the Jesuits. Of course, this plan failed.

25. The *Yijing* was popularized in the West through the translations of James Legge, *The Texts of Confucianism, Part II, The 'Yi King'* (Oxford, 1899) and Richard Wilhelm, *'I Ging'; Das Buch der Wandlungen* (Jena: Diederichs,

1924). It captured the imagination of many people from occultists such as Aleister Crowley to respected scientists, for example, C. G. Jung, founder of analytic psychology. For more information on the Chinese uses of the *Yijing*, see W. A. Sherrill and W. K. Chu, *An Anthology of I-Ching* (London: Routledge & Kegan Paul, 1977); some scientific theories of the *Yijing* are present in Z. D. Sung, *The Symbols of Yi King; or, The Symbols of the Chinese Logic of Changes* (New York: Paragon Book Reprint Corp., 1969; reprint of 1934 Shanghai edition); *Yijng* combinations are considered by Martin Gardner, "The Combinatorial Basis of the 'I Ching'; or, The Chinese Book of Divination and Wisdom," *Scientific American* 230 (1974): 108–13 and F. Van der Blij, "Combinatorial Aspects of the Hexagrams in the Chinese Book of Changes," *Scripta Mathematica* 28 (1965): 37–49.

26. Wilhelm, '*I Ging*', 504.

27. A discussion of these two cycles is given in Eva Wong, *Feng-shui: The Ancient Wisdom of Harmonious Living for Modern Times* (Boston: Shambhala, 1996), 54–56.

28. Many factors affected the diviner's response, such as the personality and status of the client and the political realities of the situation. Often the response was vague and cryptically phrased. An excerpt from *Zuo zhuan* [Master Zuo's Enlargement of the Spring and Autumn Annals] (ca. 400–250 B.C.E.) relates the advice rendered by an astrologer to Prince Ligong on the future of his son:

> The Prince asked him to tell the future of the boy by divination. He obtained the hexagram *guan* followed by the hexagram *pi* and said, "The interpretation is that *guan* represents the glory of the state and is auspicious for the guest of the Prince. Isn't this a sign that the boy will take over the state of Chen? If this is not so then he will possess another state. If this omen is not fulfilled by one of his descendants. The light extends to a far distance and will be reflected by someone else. *Kun* ☷ represents the earth; *sun* ☴ represents wind; *qian* ☰ represents heaven (From the hexagram *guan* ䷓ to the hexagram *pi* ䷋ we notice that the trigram *sun* ☴ in the former has changed to the trigram *qian* ☰ ; that is) wind has become heaven, and being above the earth, this signifies mountain. (Thus the boy) will have all the treasures of the mountains and he will be shone upon by the light of heaven. . .

29. They are: the Guardian, the Scholar, Great Gate, the Warrior, Left Guardian, Craving Wolf, Destroyer of Armies, Right Guardian and Prosperity.

30. In looking at a drawing of the *yubu* path, the concept of clockwise and counterclockwise is difficult to discern. However, subjected to a graph-theoretic interpretation where *yubu* indicates the mapping τ on the set of *luoshu* elements under the condition τ (1) = 9, a cyclic permutation in a clockwise direction is revealed. This mapping can be obtained by repeatedly adding 1 to each element of the *luoshu*, again following the condition that 9 + 1 = 1. For the counterclockwise cycle, a reverse *yubu* path is taken where 1 is repeatedly subtracted from the *luoshu* numbers with the condition that 1 − 1 = 9.

31. Information on the Chinese practices of *fengshui* reached Western audiences through the writing of the nineteenth-century missionary E. J. Eitel, *Feng Shui or the Rudiments of Natural Science in China* (Bristol: Pentacle Books, 1979; reprint of 1873 Turbner edition). He justified its description as the "science of wind and water" by noting, "it is a thing like wind which you cannot comprehend, and like water which you cannot grasp" (p. 3) and believed it was "natural science without experimentation." Eitel's definition of *fengshui* was not original and apparently came from Chinese sources. Further information on the historical background and theory of *fengshui* can be found in Wong, *Feng-shui*, 13–61 and Steven J. Bennett, "Patterns of the Sky and Earth: A Chinese Science of Applied Cosmology," *Chinese Science* 3 (1978): 1–26.

32. The Chinese calendar accommodates cycles of time and the perceived metaphysical changes that accompany the cycles. There are four basic cycles: a sexagenary cycle of sixty years; the three eras with sixty years an era; nine cycles of twenty years per cycle and twenty-four seasonal markers, two markers for each month. A listing of seasonal markers is given in Wong, *Feng-shui*, 50. Further interpretations of seasonal changes can be found in Richard J. Smith, *Chinese Almanacs* (Hong Kong: University of Hong Kong Press, 1992).

33. Instructions for constructing a geomantic chart can be found in Lillian Too, *Chinese Numerology in Feng Shui* (Kuala Lumpur: Konsep Books, 1994) and Wong, *Feng-shui*, 61–62.

34. The geomantic chart is superimposed upon a house plan with correct directional alignments and *luoshu* cell correspondences established for the rooms or parts of the house. Accumulations of negative forces in corners are to be particularly avoided. A diviner will recommend remedies such as, for example, placing a mirror to reflect forces in a particular location, as needed. See Evelyn Lip, *Feng Shui: Environments of Power: A Study of Chinese Architecture* (London: Academic Editions, 1995).

35. Yearly and monthly influences are discussed in Wong, *Feng-shui*, 217–22.

36. Numerological values as given by Too, *Chinese Numerology*, 136–53.

37. In such a situation, the fortune-teller will consider the personal dimensions and modify the predictions accordingly.

38. The "Nine Caldrons" or "Nine Tripods," as they were also called, represented the Nine Provinces of Ancient China with which the *luoshu* was sometimes associated. The caldron illustration is from *Daozang* [Daoist Patrology] (00945). Other illustrations are reproduced from Lars Berglund, *The Secret of Luo Shu: Numerology in Chinese Art and Architecture* (Lund: Lund University, 1990), 190, 166.

39. Magic diagrams are from *Daozang*.

40. For a more complete discussion of cosmic mirrors, see Schuyler Cammann, "The TLV Patterns on Cosmic Mirrors of the Han Dynasty," *Journal of the American Oriental Society* 68 (1948):159–67; Cammann, "Types of Symbols in Chinese Art," *Memoirs of the American Anthropological Association* (1953): 195–231.

41. A cosmic mirror from Wang's time bears the very commercial sounding inscription:

The Xin (dynasty) has excellent copper, it comes from Tanyang;
Refined and worked with silver and tin, it is clear and bright.
The Shang-fang [state workshops] have made (this) mirror, (which)
is completely without flaw; To the left of the Dragon and to the
right of the Tiger avert misfortune; The Vermilion Bird and the
Dark Warrior [i.e. the tortoise] are in accord with yin and yang.
May your sons and grandsons be complete in number and dwell
in the center; May you long preserve your two parents in happiness
and good fortune. May your longevity be like that of metal and
stone; May your lot be that of a prince.

Quoted in John S. Major, "The Five Phases, Magic Squares, and Schematic Cosmography," *Explorations in Early Chinese Cosmology*, ed. Henry Rosemont (Chico, CA: Scholars Press, 1984). Special issue of *Journal of the American Academy of Religious Studies*, vol. 50, 133–66.

42. Further discussion on this theory is given in Berglund, *The Secret of the Luo Shu*, 294–391.

43. See Arthur F. Wright, "The Cosmology of the Chinese City," *The City in Late Imperial China*, ed. G. William Skinner (Stanford: Stanford University Press, 1977), 33–75.

44. As given in Laurence G. Liu, *Chinese Architecture* (New York: Rizzoli, 1989), 33.

45. He Junshou, "Mingdai Beijing cheng jianzhu" [The Mathematical Base for Urban Planning of Beijing City in the Ming Dynasty], *Guang Ming Ri Bao* (August 1986). The *chi*, the Chinese foot, was equivalent to approximately 24.12 cm, a *zhang* = 10 *chi* and the *li*, the Chinese mile, spanned 1500 *chi*.

46. For a more detailed discussion of the architecture and numerology of the Forbidden City, see Evelyn Lip, *Feng Shui Environments of Power*.

Chapter 5: Chinese Variations

1. A detailed study has been made of this work: Lam Lay Yong, *A Critical Study of the Yang Hui Suan Fa: A Thirteenth-Century Chinese Mathematical Treatise* (Singapore: Singapore University Press, 1977).

2. See the discussion in chapter 2 above.

3. Schuyler Cammann, "The Magic Square of Three in Old Chinese Philosophy and Religion," *History of Religions* 1 (1961): 37–80.

4. Ibid., 55.

5. At a later period, in a different context, Daoists did seem to consider several variants of the *luoshu* in their ceremonies and as magic charms. See the discussion of "the steps of Yu" in chapter 4.

6. From earliest times, China had commercial and scientific contacts with peoples outside its borders. The Indians knew of magic squares and associated them with divination as early as 550. In the world of Islam, the *Ikhwan as-Safa* or "Brothers of Purity" published their encyclopedia *Rasa'il* in 989. It contained a section on magic squares. Most authorities believe that Yang Hui's examples were Chinese-derived.

7. In his introduction, Yang notes that he is merely passing on information he and his friends obtained from old books in their libraries. Further, he refers to squares as *yin* that are clearly *yang* in nature according to the tradition of *yinyang*. For a more complete discussion of this matter see Schuyler Cammann, "Old Chinese Magic Squares," *Sinologica* 7 (1962): 14–53.

8. His order ten square was not a true magic square—the sum of its diagonal elements did not equal the magic constant. For a full discussion on all of Yang's magic squares see Lam, *A Critical Study*, 293–311.

9. Cammann, "Old Chinese Magic Squares," 26–28.

10. A particular form of numerology popular in medieval Europe was *gematria* where every letter of a word would be associated with a number and the sum of such numbers for a word or a name was given a meaning. In Christian number mysticism, 666 was the number of the "Beast of Revelation," the devil or the Antichrist, and was associated with one's enemies or theological opponents. The process of doing this was called "Beasting." An example of how Pope Leo X was "Beasted" is related in Howard Eves, *An Introduction to the History of Mathematics* (Philadelphia: Saunders Publishing, 1990), 270.

11. That is, 15, as the smallest constant, occupies the 1's position and so on until 231, as the largest magic constant, occupies the 9's position. This feature also held for the previous square, see figure 5.15(a).

12. For a discussion of Yang's magic circles see: Lam, *A Critical Study*, 311–18, and Frank Swetz, "If the Squares Don't Get You—The Circles Will," *Mathematics Teacher* 73 (1980): 67–72.

13. The *luoshu*, itself, remained prominent as a talisman and a basis for divination and geomancy ceremonies. For a survey of Chinese work with magic

squares see, Ho Peng Yoke, "Magic Squares in China," in Selin, *Encyclopaedia of the History of Science*, 528–29.

14. *Suanfa tongzong* [Systematic Treatise on Arithmetic] gives a collection on magic squares with theoretical justifications.

15.

60	5	96	70	82	19	30	97	4	42
66	43	1	74	11	90	54	89	69	8
46	18	56	29	87	68	21	34	62	84
32	75	100	74	63	14	53	27	77	17
22	61	38	39	52	51	57	15	91	79
31	95	13	64	50	49	67	86	10	40
83	35	44	45	2	36	71	24	72	93
16	99	59	23	33	85	9	28	55	98
73	26	6	94	88	12	65	80	58	3
76	48	92	20	37	81	78	25	7	41

16. *Binaishanfang ji* [Collection of Writings in the Binai Mountain Studies].

Chapter 6: The Magic Square in Other Cultures

1. For example, the Hebrew Old Testament relates the destruction of the walls of Jericho—"For seven days, seven priests with seven trumpets invested Jericho, and on the seventh day they encompassed the city seven times." The number seven also appears in many contexts in Christian theology, including: the seven deadly sins; the seven virtues, the seven spirits of God, seven joys of the Virgin Mary, seven devils cast out of Magdalen.

2. The Tower of Babel was supposed to have been built in seven levels: one for each known planet. When King Sargon II (722–705 B.C.E.) constructed his palace at Khorsabad, he created an identity with its wall of defense by making the length of the wall 16,283 cubits, the number of his name.

3. Kabbala [Cabala, Qabalah], "the received or traditional law," was a system of mystical teaching of Judaism which became popular in Europe in the

twelfth century but which had its roots in ancient Babylonia. The teaching and practices of Kabbala were based on the belief that God created all things by pronouncing their names—words had a divine meaning and each of the twenty-two letters of the Hebrew alphabet were divine instruments. This theory was first expounded in the *Sepher Yetzirah* [Book of Creation] (ca. 3rd to 6th centuries C.E.) and expanded in the thirteenth century *Sepher Zohar* [Book of Splendor]. The task of the Kabbalist was to discover the hidden meaning in words. Gematria was a branch of Kabbala that encoded words with numbers; the hidden meanings of the numbers then reflected back on the words. Christian mystics also accepted these practices; one "theological activity" of note was to interpret the letters of an individual name so that their numerical value became 666, the number of the Beast in the Book of Revelation. For further information on these mystical number practices see Stuart Holroyd, *Magic Words and Numbers* (New York: Doubleday and Company, 1976).

4. Annmarie Schimmel, *The Mystery of Numbers* (New York: Oxford University Press, 1993), p. 215. Private consultation with independent historians and researchers in the field of ancient Babylonian mathematics supports the claim that magic squares were unknown in Babylonia.

5. Act V, Scene 1, line 2.

6. As quoted in Tobias Dantzig, *Number: the Language of Science* (New York: The Macmillan Company, 1954), 41.

7. The Pythagoreans developed a system of figurative numbers in which configurations of dots were confined to a geometric form such as a square or a triangle. The number of dots was always "square" (4, 9, 16, . . .); or "triangular" (3, 6, 10, . . .). The *tetractys* is a triangular number. For a more extensive analysis of the *tetractys* concept, see H. E. Stapleton, "Ancient and Modern Aspects of Pythagoreanism," *Osiris* 13 (1958): 12–53, especially 32–35.

8. The mathematical historian David Eugene Smith (1860–1944) noted Pythagoras's distinctly oriental ideas. See his *History of Mathematics* (New York: Dover Publications, 1958; reprint of 1925 edition).

9. Primary texts concerning Egyptian mathematics are scarce. See Frank Swetz, *From Five Fingers to Infinity: A Journey through the History of Mathematics* (Chicago: Open Court, 1994), 133.

10. The Rhind papyrus consists of a collection of eighty-five mathematical problems compiled by a scribe, Ahmes, in 1650 B.C.E. It was acquired by the Scottish antiquarian A. Henry Rhind in 1958 and, upon his death, ceded to the British museum.

11. Takao Hayashi, "Magic Squares in Indian Mathematics," *Encyclopedia of the History of Science, Technology, and Medicine in Non-Western Cultures*, ed. Helaine Selin (Dordrecht: Kluwer Academic Publishers, 1997), 529–36.

12. Bibhutibhusan Datta and Awadhesh Narayan Singh, "Magic Squares in India," *Indian Journal of Science* 27 (1992): 51–120.

13. Schuyler Cammann, "Islamic and Indian Magic Squares, Part II," *History of Religions* 8 (1969): 271–99.

14. These squares are given in Takanori Kusuba, "Combinatories and Magic Squares in India: A Study of Narayana Pandita's *Ganitakaumudi*," chaps. 13–14. Ph.D. dissertation, Department of History, Brown University, 1993, p. 169. The ancient Indians conceived of nine planets: the "Seven Luminaries" or Heavenly Wanders plus two imaginary invisible planets, Rahu and Ketu. These nine magic squares of order three can be obtained from the *luoshu* by rotations and augmentations, namely by adding the same constant to all entries in a given magic square.

15. Hayashi, "Magic Squares in Indian Mathematics," 169.

16. See Arion Rosu, "Etudes ayurvediques III Les carrés magiques dans la médecine indienne," *Studies on Indian Medical History*, G. Jan Meulenbeld and Dominik Wujastyk (Groningen: Egbert Forsten, 1987), 103–12.

17. For a review of Frost's work see "Indian Magic Squares," in Edward Falkener, *Games Ancient and Oriental and How to Play Them* (New York: Dover Publications, 1961; reprint of Longmans, Green edition, 1892).

18. See Hiralal R. Kapadia, "A Note on Jaina Hymns and Magic Squares," *Indian Historical Quarterly* 10 (1934): 140–53.

19. See Kusuba, "Combinatorics and Magic Squares in India."

20. Aagar is a type of wood found in India. These directions are provided in L. R. Chawdhri, *Practicals of Yantras* (New Delhi: Sagar Publications, 1984), 35.

21. Ibid., 47.

22. Captain Shortreede, "On an Ancient Indian Magic Square, Cut in a Temple at Gwalior," *Journal of the Asiatic Society of Bengal* 11 (1842): 292.

23. Cammann, "Islamic and Indian Magic Squares, Part II," 275.

24. See Yukeo Ohashi, "Astronomy in Tibet," in Selin, *Encyclopaedia of the History of Science,* 136–39. The Shixian calendar was the last luni-solar calendar devised in China.

25. The almanacs feature the *luoshu* as the centerpiece which is usually surrounded by the twelve animals the Chinese use to designate each year in their twelve year cycle: rat, ox, tiger, rabbit, dragon, snake, horse, sheep, monkey, bird, dog and boar. This configuration is depicted on the undershell of a ferocious looking tortoise. Figure 6.9 is a design on a silver amulet as shown in Antoinnette K. Gordon, *Tibetan Religious Art* (New York: Paragon, 1963), 89.

26. For example, *Kigu hosu* (1697) by Yueki Ando.

27. See "Magic Square in Japanese Mathematics" by Yoshimasa Michiwaki in Selin, *Encyclopaedia of the History of Science*, 538–40.

28. Ibid., 538.

29. *Shochu shinan* (1669). I am grateful to Yoshimasa Michiwaki, President of Maebashi Institute of Technology, Maebashi, Japan for sending me these materials.

30. See William Ahrens, "Studien über der magischen Quadrate der Araber," *Der Islam* 7 (1917): 186–249. This article still remains the most comprehensive text on magic squares in Islamic written in a Western language.

31. Jabirian numerology is discussed by: H. E. Stapleton, "The Antiquity of Alchemy," *Ambix* 5 (1953): 1–43; Syed Nomanul Haq, *Names, Natures and Things: The Alchemist Jabir ibn Hayyan and his Kitab al-Ahjar* (Boston: Kluwer Academic Publishers, 1994).

32. See H. E. Stapleton, "The Gnomon," *Ambix* 6 (1957): 1–9.

33. As discussed in Seyyed Hossein Nasr, *An Introduction to Islamic Cosmological Doctrines* (Albany: SUNY Press, 1993).

34. See Georges Ifrah, *From One to Zero: A Universal History of Numbers* (New York: Penguin Books, 1985), chap. 20: "Arabic Numeral Letters."

35. The connection between magic squares and alchemy is discussed in Vladimir Karpenko: "Between Magic and Science: Numerical Magic Squares," *Ambix* 40 (November, 1993): 121–28; "Two Thousand Years of Numerical Magic Squares," *Endeavor* 18 (1994): 147–53. I thank Jacques Sesiano for calling my attention to these articles.

36. For more details see, Cheng Te-k'un, "Some Chinese Islamic Magic Square Porcelain," *Wen wu hui k'an* 1 (1972): 146–60.

37. See Schuyler Cammann, "Islamic and Indian Magic Squares, Part I," *History of Religions* 8 (1969): 181–209.

38. Nasr, *An Introduction to Cosmological Doctrines*, p. 50. For a more detailed discussion of Ikhwanian beliefs see Seyyed Hossein Nasr, *Science and Civilization in Islam* (Cambridge, MA: Harvard University Press, 1968), 152–57.

39. As given in B. Carra de Vaux, *Les Penseurs de l'Islam*, vol. 4 (Paris: Geuthner, 1923), 109–10.

40. As described by Cammann, "Islamic and Indian Magic Squares, Part I," 194.

41. See John Stroyls, "Survey of the Arab Contributions to the Theory of Numbers," *Proceedings of the First International Symposium for the History of Arabic Sciences*, vol. 2, Paper in European Languages (Aleppo: Institute for the History of Arabic Sciences, April 5-12, 1976), 168–79.

42. Stapleton, "Antiquity," p. 11.

43. For example, the Persian manuscript from 1212, Princeton University Library, Garrett Collection No. 1057; British Museum MS (Add. 7713), dates at 1211.

44. The torus only became an entity of mathematical interest in seventeenth-century Europe.

45. See Ifrah, *From One to Zero*.

46. Nasr, *Science and Civilization in Islam*, p. 210.

47. I would like to thank Mohammad Bagheri, Director, Encyclopedia Islamica Foundation, Tehran, for sending information on "Beduh" and my colleagues, Ali Behagi and Rafik Culpan, for their assistance in translating Farsi and Turkish language materials.

48. Photos of such children are given in W. Ahrens and Alfred Maass, *Etwas von magischen Quadraten in Sumatra und Celebes* (Berlin, 1916).

49. Walter Skeat, *Malay Magic* (New York: Dover Publications, 1967).

50. Ibid., 138.

51. Edward Westermarck, *Ritual and Belief in Morocco,* vol. 1 (London: Macmillan Co., 1926), 145.

52. As quoted in Claudia Zaslavsky, *Africa Counts: Number and Pattern in African Culture* (Boston: Prindle, Weber & Schmidt 1973), 139.

53. The "Four Angels" are most probably: the archangels, Gabriel and Michael; Izra'il, the Angel of Death, and Malik, the gatekeeper of hell. The Quran only mentions six angels by name; the remaining two are Harut and Marut who are known for prompting demons to cause mischief.

54. Edward Lane, *An Account of the Manners and Customs of the Modern Egyptians* (London, 1836).

55. Incident related in Ahrens, "Studien über der magischen Quadrate der Araber," 224.

56. Illustration given in Ifrah, *From One to Zero,* 309.

57. This square is given in Abd Al-Fattah Al-Sayyid Al-Tukhi, *Kitab Sihr Al-Kuhhan, Fi Hudur Al-Jann* [The Secrets of Magicians in their Intercourse with Spirits], date and place of origin unknown. The book was purchased in Egypt in 1988, see Berglund, *The Secret of the Luo Shu,* 52–55.

58. Table given in Stapleton, "Antiquity," 24.

59. See Emilia Calvo, "Al-Majriti," in Selin, *Encyclopaedia of the History of Sciences,* 547.

60. See Menso Folkerts, "Zur Frühgeschichte der magischen Quadrate in Westeuropa," *Sudhoffs Archiv* 4 (1981): 313–38.

61. Jagiellonian Library, MS 753.

62. Folkerts, "Zur Frühgeschichte der magischen Quadrate in Westeuropa," discusses early codices and manuscripts that contain information on magic squares.

63. A survey of Abraham ben Meir Ezra's life and works can be found in Raphael Levy, *The Astrological Works of Abraham ibn Ezra,* vol. III (Baltimore, MD: The Johns Hopkins University Press, 1927).

64. The nature of his astrological considerations is revealed in Raphael Levy and Francisco Cantera, *The Beginning of Wisdom: An Astrological Treatise by Abraham ibn Ezra,* vol. XIV (Baltimore, MD: The Johns Hopkins University Press, 1939).

65. M. de la Hire's translation of Moschopoulos's work was never published. In 1886, Paul Tannery rendered it into French as *"Le traite de Manuel Moschopoulos sur les carres magiques,"* later translated and published in English: John Calvin McCoy, "Manuel Moschopoulos's Treatise on Magic Squares," *Scripta Mathematica* 8 (1941): 15–26.

66. Kurt Vogel, *Die Practica des Algorismus Ratisbonensis* (Munich: C. H. Beck, 1954).

67. David Singmaster, "Fair Division of the First kn Integers into k Parts," prepublication draft received through private correspondence.

68. Ibid.

69. Problems 12 and 51 in the collection. See David Singmaster and John Hadley, "Problems to Sharpen the Young," *The Mathematical Gazette* (March, 1992): 102–26.

70. Folkert, "Zur Frühgeschichte der magischen Quadrate in Westeuropa," 321.

71. The topic and effects of melancholy was the subject of much medieval speculation and artistic and literary attention. See Raymond Kilbansky, Erwin Panofsky and Fritz Saxl, *Saturn and Melancholy: Studies in the History of Natural Philosophy, Religion and Art* (New York: Basic Books, 1964).

72. Much analysis has been devoted to the symbolic significance of the items depicted in Dürer's *Melancholia I*. See, for example, Edwin Panofsky, *Albrecht Dürer* (Princeton, Princeton University Press, 1943). The prevailing explanation for the presence of the magic square of order five is that it was to counter Jupiter's malevant power; however, Cammann believes it is merely to represent mathematics among the other sciences depicted (Cammann, "Islamic and Indian Magic Squares," 292).

73. Scipione del Ferro, Professor of Mathematics at the University of Bologna, is perhaps best known for his contributions to the solution of the cubic equation. See discussion in Victor J. Katz, *A History of Mathematics* (New York: Harper Collins, 1993), 328–30.

74. The contents of this table are adapted from Donald Tyson, *Three Books of Occult Philosophy written by Henry Cornelius Agrippa* (St. Paul, MN: Llewellyn Publications, 1995), Appendix VII, "Practical Kabbalah," 763.

75. See discussion in T. Schrire, *Hebrew Amulets: Their Decipherment and Interpretation* (London: Routledge & Kegan Paul, 1966), chap. 17, "Construction of the Shemoth," 91–97.

76. Tyson, *Three Books of Occult Philosophy*, 766–69.

77. Schrire, *Hebrew Amulets*, 403–5.

78. Tyson, *Three Books of Occult Philosophy*, translation by James Freake.

79. In occult practice, "seals" were symbols that represented the object of attention or some of its special properties. In his collection of magic squares, Agrippa usually presents three seals for each planet: the seal of the planet; the seal of its "intelligence"; and the seal of its "spirit."

80. Tyson, *Three Books of Occult Philosophy*, 318.

81. Other users of this system also reversed the order. See the discussion on this matter in Folkert, "Zur Frühgeschichte der magischen Quadrate in Westeuropa," 315.

82. *Du Royaume de Siam* by Simon de la Loubère was published in French in Paris (1691) and Amsterdam (1713) and appeared in English translation in 1693. Europeans at this time were fascinated with accounts concerning the civilizations of the East. It was this book that originally aroused my interest in the *luoshu*.

83. The question, "How many magic squares of order four were possible?," plagued the Arabs. In the period from 1837 to 1838, Violle published a three-volume study of magic squares, *Traite complet des Carres Magiques*. In it, he noted that there existed 549,504 different numerical arrangements for a fifth-order magic square.

84. Franklin's work with magic squares is referred to in William B. Willcox, ed., *The Papers of Benjamin Franklin* (New Haven: Yale University Press, 1972). In vol. 15, pp. 171–72, a letter of July 2, 1768 from Franklin to John Winthrop notes that the British are puzzled about magic squares but they have not "asked to see my methods." In Franklin's correspondence are several references to magic squares.

85. In particular Frost wrote a pioneering article on magic squares for *Encyclopaedia Britannica* (1882) which was used in later editions. This article introduced the English-speaking world to the mystery and fascination of magic squares.

Chapter 7: *Luoshu* Miscellanea

1. Richard Webster, *Talisman Magic: Yantra Squares for Tantric Divination* (St. Paul, MN: Llewellyn Publications, 1995), preface. James Ward, in his article, "Vector Spaces of Magic Squares," *Mathematics Magazine* 53 (March 1980): 108–11, also mentions this by quoting Hutton's *Mathematical Recreations* (1844):

> According to this idea a square of one cell filled up with unity, was the symbol of the Deity, on account of the unity and immutability of God; for they remarked that this square was, by its nature, unique and immutable, the product of our unity by itself being always unity. The square of the root two was the symbol of imperfect matter, both on account of the four elements and the impossibility of arranging this square magically . . .

2. Gino Loria, *Le Scienze esatte nell'Antica Grecia* (Milan, 1914), 795, suggested that Theon's square was a magic square and this belief was held by later authors. For example, James Moran, *The Wonders of Magic Squares* (New York: Random House, 1982), 5.

3. This proof is expanded upon in William H. Benson and Oswald Jacoby, *New Recreations with Magic Squares* (New York: Dover Publications, 1976), 104–7.

4. R. Holmes, "The Magic Magic Square," *The Mathematical Gazette* (December 1970): 376.

5. Hwa Suk Hahn, "Another Property of Magic Squares," *The Fibonacci Quarterly* 73 (1975): 205–8. I would like to thank Martin Gardner for calling my attention to this article.

6. For further details on such squares, see Martin Gardner. "Some New Discoveries About 3 × 3 Magic Squares," *Math Horizons* (February 1998): 11–13.

7. Ibid.

8. For a clearer concept of "vector space," I recommend consulting a book on linear algebra, for example: Gilbert Strang, *Linear Algebra and its Applications* (New York: Academic Press, 1976).

9. See Martin Cohen and John Bernard, "From Magic Squares to Vector Spaces," *Mathematics Teacher* (January 1982): 76–77, 64; C. Small, "Magic Squares over Fields," *American Mathematical Monthly* 95 (1988): 621–25; A. van den Essen, "Magic Squares and Linear Algebra," *American Mathematical Monthly* 97 (1990): 60–62; and Ward, "Vector Spaces of Magic Squares."

10. Scholars disagree on the exact definition of a magic square. M. M. Postinikov, writing in the *Soviet Mathematical Encyclopaedia*, vol. 6 (Kluwer, 1990) 72, defines a magic square as an "n × n array of integers from 1 up to n^2 . . ."; Sherman Stein, in *Mathematics: The Man-Made Universe* (San Francisco: W. H. Freeman, 1963, p. 167), specifies "an arrangement of n^2 natural numbers from 1 to n^2. . ."; Howard Eves, in *An Introduction to the History of mathematics*, 6th edition (Philadelphia: Saunders College Publishing), says all authorities agree on the magic sum property. In general, the squares we have defined and which Eves speaks of are frequently referred to as "classical" magic squares.

11. Matrix multiplication on two three-by-three matrices is performed as follows:

$$\begin{bmatrix} a_{11}, a_{12}, a_{13} \\ a_{21}, a_{22}, a_{23} \\ a_{31}, a_{32}, a_{33} \end{bmatrix} \begin{bmatrix} b_{11}, b_{12}, b_{13} \\ b_{21}, b_{22}, b_{23} \\ b_{31}, b_{32}, b_{33} \end{bmatrix} \begin{bmatrix} a_{11}b_{11} + a_{12}b_{21} + a_{13}b_{31}, & a_{11}b_{12} + a_{12}b_{22} + a_{13}b_{32}, & a_{11}b_{13} + a_{12}b_{23} + a_{13}b_{33} \\ a_{21}b_{11} + a_{22}b_{21} + a_{23}b_{31}, & a_{21}b_{12} + a_{22}b_{22} + a_{23}b_{32}, & a_{21}b_{13} + a_{22}b_{23} + a_{23}b_{33} \\ a_{31}b_{11} + a_{32}b_{21} + a_{33}b_{31}, & a_{31}b_{12} + a_{32}b_{22} + a_{33}b_{32}, & a_{31}b_{13} + a_{32}b_{23} + a_{33}b_{33} \end{bmatrix}$$

12. A semi-magic square is a square in which one or both of the main diagonals do not total to the magic sum. See Emanuel Emanouilidis, "Powers of Magic Squares," *Journal of Recreational Mathematics* 29 (1998): 176–77.

13. For example: C. W. Trigg, "Magic Square as a Determinant," problem response E813, *American Mathematical Monthly* 56 (January 1949): 33–37; A. C. Thompson, "Odd Magic Powers," *American Mathematical Monthly* 101 (April 1994): 339–42; David Rose, "Magic Square and Matrices," *The Mathematical Gazette* (February 1973): 36–39; John Robertson, "Magic Squares of Squares," *Mathematics Magazine* 69 (October 1996): 289–93.

14. As related in Menso Folkerts, "Zur Frühgeschichte der magischen Quadrate in Westeuropa," *Südhoffs Archiv* 65 (1981): 313–38.

15. Henry Ernest Dudeney, *536 Puzzles & Curious Problems* (New York: Charles Scribner's Sons, 1967), 386, problem 142.

16. This is problem 545 submitted by G. W. Walker. The problem was solved by C. W. Trigg, "The Mathematician and the Jester, " *American Mathematical Monthly* 55 (September 1948): 429–30.

17. Ann-Lee Wang, "Hollow Magic Squares," *Mathematics in School* (March 1995): 23–25; also see, Betty Lyon, "Using Magic Borders to Generate Magic Squares," *Mathematics Teacher* (March 1984): 223–26.

18. Wang, "Hollow Magic Squares," 23.

19. Martin Gardner, *The Numerology of Dr. Matrix* (New York: Simon and Schuster, 1967), 89. Loyd published his in 1928.

20. The concept of a heterosquare was first proposed by Dewey Duncan in the "Problems and Solutions" section of *Mathematics Magazine* (January 1951) where the question of a heterosquare of order three was raised. Solutions for this question were given by Charles Pinzka, "Heterosquares," problem 84, *Mathematics Magazine* (September/October 1965): 250–52 and by Charles W. Trigg, "Comment on Problem 84," *Mathematics Magazine* (September/October 1971): 236–37.

21. As given in Martin Gardner, "The Magic of 3×3," *Quantum* 6 (January/February 1996): 24–28.

22. Ibid., p. 25.

23. See Robertson, "Magic Squares of Squares." A discussion of Fermat's Last Theorem is given in Charles Vanden Eynden, "Fermat's Last Theorem" and Frank J. Swetz, "Epilogue: Fermat's Last Theorem" in *From Five Fingers to Infinity: A Journey through the History of Mathematics*, ed. Frank J. Swetz (Chicago: Open Court, 1994), 747–50, 751–52, respectively.

24. Given in Underwood Dudley, *Numerology, or, What Pythagoras Wrought* (Washington, DC: Mathematical Association of America, 1997), 68.

25. Gardner, *Numerology*, 49.

26. For a more complete discussion of magic cubes, see William Symes Andrews, *Magic Squares and Cubes* (New York: Dover Publications, 1960; reprint of Open Court edition, 1917).

27. See Lee C. F. Sallows, "Alphamagic Squares: Adventures with Turtle Shell and Yew between the Lowlands of Logology," *Abacus* (Fall 1986): 28–45, and (Winter 1987): 20–29, 43.

28. Ibid., 41, 43, 44–45.

29. Ibid., 31.

30. Andrews, *Magic Squares and Cubes*, 146–58.

31. As quoted in Andrews, ibid., 148–49.

32. Communicated to the author in private correspondence.

33. Michael R. Saso, in his *Taoism and the Rite of Cosmic Renewal* (Pullman, WA: Washington State University Press, 1972), describes a *yubu* dance for the Chiao festival of cosmic renewal, p. 71.

34. David Roberts, "Peter Maxwell Davies: *Ave Maris Stella*," *Contact* 19 (1978): 26–31.

35. Claude Bragdon, *The Frozen Fountain Being: Essays on Architecture and the Art of Design in Space* (New York: Books for Libraries Press, 1970; reprint of 1924 edition).

36. Ibid., 44, 76.

37. Ibid., 76.

38. Ibid., 83.

39. A general technique was suggested in Margaret J. Kenney, "An Art-Full Application Using Magic Squares," *Mathematics Teacher* 75 (January 1982): 83–89; it was used with the *luoshu* by Roger Enge, "Reader Reflections: Magic-Square Designs," *Mathematics Teacher* 77 (November 1984): 596.

40. These possibilities are discussed in Rebecca Bloxham, "Patterns within Patterns: The Fractal Nature of Ancient Chinese Number Patterns," paper presented at the Fifth International Conference on the History of Science in China, University of California, San Diego (August 5–10, 1988). Ms. Bloxham has also developed a lithograph process which has resulted in the creation of intriguing patterns. She has named her process after the *luoshu*. See Rebecca Bloxham, "Luo-Shu Washes," *The Tamarind Papers* 6 (1982): 22–24.

41. Ibid., 13.

42. As translated and quoted in Bradford Tyrey and Marcus Brinkman, "The *Luo-shu* as *Taiji* Boxing's Secret Inner-Sanctum Training Method," *Journal of Asian Martial Arts* 5 (1996): 75–79.

43. Ibid., 77–78.

44. Bow-Sim Mark, *Simplified Tai-Chi Chuan* (Boston: Chinese Wushu Research Institute, 1982), 22.

Chapter 8: Some Final Thoughts

1. In "Aspects of Pythagorism" and "Antiquity of Alchemy," H. E. Stapleton credits Apollonius of Tyana and Theodorus of Asine with the creation of magic squares, but does not explain his reasons. For a discussion of Theon of Smyrna's contributions b to magic squares, see p. 122 above.

2. This historical aspect of magic squares has been thoroughly researched and documented by Jerome Carcopino, "Le Christianisme secret du 'carré magique'," *Museum helveticum* 5 (1948): 16–59.

3. While each of the words, with the exception of *arepo*, which is a proper name, have a meaning within a Latin context—*rotas*, 'wheels'; *opera*, 'labor or

work'; *tenet*, 'to hold', and *sator*, 'sower of seeds'—in the context of the square, they make no sense.

4. Carcopino, "Le Christianisme," 33, 35. In Revelation I: 8, St. John has Christ proclaiming, "I am the Alpha and the Omega, the beginning and the end." Alpha and omega are the first and last letters of the Greek alphabet. Revelations was written in Greek for a Greek audience where the symbolism would be obvious.

5. A discussion of the occult uses of the SATOR square is given in Stuart Holroyd, *Magic, Words and Numbers* (New York: Doubleday & Co., 1976), 80–81.

6. Berglund, *The Secret of Luo Shu*, has devoted a section of his work to "Similarities between Chinese and Greek numerology," 48–50.

7. Joseph Needham, *Science and Civilisation in China*, vol. 5: *Chemistry and Chemical Technology* (Cambridge: Cambridge University Press, 1980), 463.

8. For a fuller discussion of early Chinese counting board practices, see Lam and Ang, *Fleeting Footsteps*.

9. This symbol evolved into the Seal of Saturn, as given in figure 6.30 above.

10. James Legge, *The I-Ching* (New York: Dover Publications, 1963; reprint of 1899 edition), 18.

11. Granet, *La Pensée chinoise*.

12. Hermann Schubert, *Mathematical Essays and Recreations* (Chicago: Open Court Publishing Co., 1903), 39, 41.

13. Martin Gardner, *Penrose Tiles to Trapdoor Ciphers* (New York: W. H. Freeman, 1989), 296.

14. The mathematical and philosophical significance of the *sriyantra* is discussed in George Gheverghese Joseph, *The Crest of the Peacock* (London: I. B. Tauris, 1991), 238–41. The mandala as a visual stimulus to contemplation is also considered in Robert Lawlor, *Sacred Geometry* (London: Thames and Hudson, 1982).

15. The east, "Place of the Dawn," had the color yellow and was thought to be "fertile and good"; the north, "Region of the Underworld," had the color red and was "barren" and "bad"; west, the "Region of Women," was blue-green and was considered "unfavorable" and humid; the south was the "Region of Thorns" and bore the color white.

16. Granet, "Les Nombres," *La Pensée chinoise*, 173–74.

17. Kepler's intellectual and scientific ordeal is examined in Arthur Koestler, *The Sleepwalkers* (New York: Grosset & Dunlap, 1963), part 4.

18. Benjamin Franklin, in describing one of the magic squares he created, noted that it is "the most magical of any magic square made by any magician"; quoted in Pieter van Delft and Jack Botermans, *Creative Puzzles of the World* (New York: Harry N. Abrams Inc., 1978), 86.

Epilogue

1. Cuneiform texts from the Seleucid period of Babylonia history note five celestial bodies, identified by the naked eye as moving through the night sky along an identifiable path. We recognize these bodies as the planets Jupiter, Venus, Mercury, Saturn, and Mars. Adding the sun and the moon, we arrive at the "seven heavenly wanderers" believed by some ancient peoples to be gods. Based on their periods of sidereal rotation, the Greeks ordered the "wanderers" as follows: Sun, Moon, Mars, Mercury, Jupiter, Venus, and Saturn, with each serving as a "ruler" for one of seven consecutive days. This sequence determined the days of the week as followed in the Western world. See O. Neugebauer, *The Exact Sciences in Antiquity* (New York: Dover Publications, 1969; reprint of 1957 edition), 168–69.

2. For further discussion of the ritual origins of science, see Abraham Seidenberg, "The Ritual Origin of Science," *Archive for History of Exact Sciences* 1 (1962): 488–527.

3. This theory is developed in Frank Swetz, "Trigonometry Comes Out of the Shadows," *Learn from the Masters*, ed. Frank Swetz et al. (Washington, DC: Mathematical Association of America, 1995), 57–71.

4. For information on Stonehenge, see Gerald Hawkins, *Stonehenge Decoded* (New York: Doubleday, 1965); for information on the Sun Dagger, see Anna Sofar, *The Sun Dagger* (Bethesda, MD: Atlas Video, 1993). Video recording.

5. The judgment of just what comprises "scientific thinking" is relative. All traditional peoples have some form of science, an ethnoscience. In the context above, "prescientific thinking" is that which takes places outside of the Western deductive-based paradigm founded on the scientific method consisting of observation, conjecture of hypothesis, experimentation, and conclusion.

6. Actually the Greeks conjectured that the planets moved around the Earth in circular orbits with each orbit contained within a crystal sphere. Thus the Earth itself was encased in a series of seven crystalline spheres that vibrated due to their planets' rotation. If the planets rotated at their proper frequency a harmonic resonance would be established—the "music of the spheres." However, if the heavenly bodies did not perform correctly, a dissonance would result causing celestial and human tension.

7. In particular René Descartes (1596–1650), French philosopher and natural scientist, used the example of a circle in his metaphysical and theological discussions. Gottfried Wilhelm Leibniz (1646–1716), Newton's rival in the seventeenth century, also held such a view. His circular conception of God was influenced by the theories of the German mystic, Nicholas of Cusa (1401–1464), who believed that God could only be appreciated by mystical intuition.

8. For a consideration of geometric shape and culture, see Robert Lawlor, *Sacred Geometry* (London: Thames and Hudson, 1982).

9. The literature on the mystical power of numbers is extensive. See, for example: W. Wynn Westcott, *Numbers, Their Occult Powers and Mystic Virtues* (London: Theosophical Publishing Society, 1911) and Christopher Butler, *Number Symbolism*, (London: Routledge and Kegan Paul, 1970).

10. Think of the football chant so often used by fans: "We're number one!"

11. For many people, the number thirteen is considered an unlucky number. This belief goes back at least two thousand years to ancient Babylonia where it was felt twelve was a good number for astrological and mathematical reasons. The rotation of the night sky was divided into twelve parts. Thus a complete cycle encompassed twelve units, thirteen, one beyond twelve, was considered undesirable or unlucky.

Bibliography

Aero, Rita. *Things Chinese*. New York: Doubleday, 1980.
Ahrens, Wilhelm. "Studien über de magischen Quadrate der Araber." *Der Islam* 7 (1917): 186–250.
Ahrens, Wilhelm, and Alfred Maass. *Etwas von magischen Quadraten in Sumatra und Celebes*. Berlin: 1916.
Allen, Sarah. *The Way of Water and Sprouts of Virtue*. Albany, NY: SUNY Press, 1997.
———. *The Shape of the Turtle: Myth, Art, and Cosmos in Early China*. Albany, NY: SUNY Press, 1991.
Andersen, Poul. "The Practice of Bugang." *Cahiers d'Exteme-Asie* 5 (1989-90): 15–53.
Andrews, William. *Magic Squares and Cubes*. New York: Dover Publications, 1960. Reprint of 1917 edition.
Benjamin, Arthur T., and Kan Yasuda. "Magic 'Squares' Indeed!" *American Mathematical Monthly* 106 (February 1999): 152–56.
Bennett, Steven. "Patterns of Sky and Earth: A Chinese Science of Applied Cosmology." *Chinese Science* 3 (1978):1–26.
———. "Chinese Science: Theory and Practice." *Philosophy East and West* 28 (1978): 439–53.
Benson, William, and Oswald Jacoby. *New Recreations with Magic Squares*. New York: Dover Publications, 1976.
Berglund, Lars. *The Secret of Luo Shu: Numerology in Chinese Art and Architecture*. Lund: Lund University, 1990.
Bergsträsser, G. "Zur den magischen Quadraten." *Der Islam* 13 (1923): 227–35.
Biggs, N. L. "The Roots of Combinatorics." *Historia Mathematica* 2 (1979): 109–36.
Blacker, Carmen, and Michael Loewe, eds. *Ancient Cosmologies*. London: George Allen & Unwin, 1975.

Blau, Joseph. *The Christian Interpretation of the Cabala in the Renaissance.* New York: Columbia University Press, 1944.

van der Blij, F. "Combinatorial Aspects of the Hexagrams in the Chinese Book of Changes." *Scripta Mathematica* 28 (1965): 37–49.

Blofeld, John. *The Secret and the Sublime: Taoist Mysteries and Magic.* London: George Allen & Unwin, 1973.

Bloxham, Rebecca. "Patterns within Patterns: The Fractal Nature of Ancient Chinese Number Patterns." Paper presented at the Fifth International Conference on the History of Science in China, University of California, San Diego, August 5–10, 1988.

———. "Luo-Shu Washes." *The Tamarind Papers* 6 (1982): 22–24.

Bodde, Dirk. *Chinese Thought, Science and Society.* Honolulu: University of Hawaii Press, 1991.

———. "Types of Chinese Categorical Thinking." *Journal of the American Oriental Society* 59 (1939): 201–21.

Bow-Sim Mark. *Simplified Tai-Chi Chuan.* Boston: Chinese Wushu Research Institute, 1982. 22.

Boyile, Veolia J. *The Fundamental Principles of the Yi-king, Tao and the Cabbalas of Egypt and the Hebrews.* London: W & G Foyle, 1934.

Bragdon, Claude. *The Frozen Fountain Being: Essays on Architecture and the Art of Design in Space.* New York: Books for Libraries Press, 1970.

———. *The Arch Lectures.* New York: Creative Age Press, 1942.

———. *Old Lamps for New: The Ancient Wisdom of the Modern World.* New York: Alfred Knopf, 1925.

Budge, E. A. Wallis. *Amulets and Talismans.* New Hyde Park, NY: University Books, 1961.

Burckhardt, Titus. *Sacred Art in East and West.* Pates Manor, UK: Perennial Books, 1967.

Burkert, Walter. *Lore and Science in Ancient Pythagoreanism.* Cambridge, MA: Harvard University Press, 1972.

Burnett, J. C. *Easy Methods for Constructing Magic Squares.* London: Rider & Co., 1936.

Butler, Christopher. *Number Symbolism.* London: Routledge & Kegan Paul, 1970.

Calder, I. R. F. "Magic Squares in the Philosophy of Agrippa of Nettlesheim." *Journal of Warburg and Courtauld Institute* 12 (1949): 196–99.

Cammann, Schuyler. "Islamic and Indian Magic Squares." *History of Religions* 8 (1969): 181–209; 271–99.

———. "Old Chinese Magic Squares." *Sinologica* 7 (1962): 14–53.

———. "The Magic Square of Three in Old Chinese Philosophy and Religion." *History of Religions* 1 (1961): 37–80.

———. "The Evolution of Magic Squares in China." *American Oriental Society Journal* 80 (1960): 116–24.

———. "Types of Symbols in Chinese Art." *Memoirs of the American Anthropological Association* (1953): 195–231.

———. "The TLV Pattern on Cosmic Mirrors of the Han Dynasty." *Journal of the American Oriental Society* 68 (1948): 159–67.
Carcopino, Jérôme. *Les Fouilles de Saint-Pierre et La Tradition le Christianisme Secret du Carré Magique*. Paris: Albin Michel, 1953.
———. "Le Christianisme secret du carré magique." *Museum helveticum* 5 (1948): 16–59.
Carra de Vaux, B. *Les Penseurs de l'Islam*, vol. 4. Paris: Geuthner, 1923. 109–10.
Carus, Paul. *Chinese Astrology*. Chicago: Open Court, 1974. Reprint of 1907 *Chinese Thought*.
———. "Chinese Philosophy." *The Monist* 6 (1896): 188–249.
Cassirer, Ernst. *The Philosophy of Symbolic Forms*. 2 vols. New Haven: Yale University Press, 1955.
Cavendish, Richard, ed. *Encyclopedia of the Unexplained: Magic, Occultism and Parapsychology*. New York: McGraw Hill, 1974.
———. *Man, Myth and Magic: An Illustrated Encyclopedia of the Supernatural*. New York: Cavendish Corp and London BPC Publishing, 1970.
Chang, Chao-Kang. *China: Tao in Architecture*. Boston: Berkhauser, 1987.
Chaudhary, C. V. *Vedic Numerology*. Bombay: Bharatiya Vidya Bhavan, 1968.
Chawdhri, L. R. *Practicals of Yantras*. New Delhi: Sagar Publications, 1984.
Cheng, David. "On the Mathematical Significance of the Chinese Ho T'u and Lo Shu." *American Mathematical Monthly* 32 (1925): 499–504.
Cheng, Te-k'un. "Some Chinese Islamic Magic Square Porcelain." *Journal of Asian Art* 1 (1974): 146–61.
Chu, W. K. *The Astrology of the I-Ching*. Edited by W. A. Sherrill. New York: Samuel Weiser, 1980.
Chuen, Lam Kan. *Feng Shui Handbook*. New York: Henry Holt, 1996.
Cirlot, J. E. *A Dictionary of Symbols*. New York: Philosophical Library, 1971.
Clarke, J. J. *Oriental Enlightment: The Encounter Between Asian and Western Thought*. London: Routledge, 1977.
Clayre, Alasdair. *The Heart of the Dragon*. Boston: Houghton Mifflin Co., 1985.
Clulee, Nicholas H. *John Dee's Natural Philosophy: Between Science and Religion*. London: Routledge, 1988.
Cohen, Martin P., and John Bernard. "From Magic Squares to Vector Spaces." *Mathematics Teacher* (1982): 76–77.
von Collani, Claudia. *P. Joachim Bouvet S. J.: Sein Leben und sein Werk*. *Monumenta Serica* monograph series, no. xvii. Nettetal: Steyler Verlag, 1985.
Colville, W. J. *Kabbalah, the Harmony of Opposite*. New York: Macoy Publisher & Masonic Supply Co., 1916.
Comber, L. *Chinese Magic and Superstitution in Malaysia*. Singapore: D. Moore for Eastern Universities Press, 1960.
Contenau, Georges. *Everyday Life in Babylonia and Assyria*. New York: W. W. Norton & Co., 1966.

Cook, Daniel J., and Henry Rosemont, Jr. "The Pre-established Harmony Between Leibniz and Chinese Thought." *Journal of the History of Ideas* 42 (1981): 3.

Cooper, J. C. *An Illustrated Encyclopedia of Traditional Symbols*. London: Thames and Hudson, 1978.

Coudert, Allison P. *Leibniz and the Kabbalah*. Boston: Kluwer Academic Publishers, 1995.

Critchlow, Keith. *Islamic Patterns: An Analytic and Cosmological Approach*. London: Thames and Hudson, 1976.

Datta, H., and A. N. Singh. "Magic Squares in India." *Indian Journal of the History of Science* 27 (1992): 51–120.

Dantzig, Tobias. *Number: the Language of Science*. New York: Macmillan Company, 1954.

David-Neel, Alexandria. *Magic and Mystery in Tibet*. New York: Dover Publications, 1971. Translation of 1929 edition.

De Bary, William. Theodore, et al. *Sources of Chinese Tradition*. 2 vols. New York: Columbia University Press, 1964.

van Delft, Pieter, and Jack Botermans. *Creative Puzzles of the World*. New York: Harry N. Abrams Inc., 1978.

Diana, Lind Mae. "The Peruvian Quipu." *Mathematics Teacher* 60 (Oct. 1967): 623–28. Reprinted in *From Five Fingers to Infinity*, ed. Frank Swetz. Chicago: Open Court, 1994. 80–85.

Doeringer, Franklin M. "The Gate in the Circle: A Paradigmatic Symbol in Early Chinese Cosmology." *Philosophy East and West* 32 (1982): 309–24.

Douglas, Nik. *Tibetan Tantric Charms and Amulets*. New York: Dover Publications, 1978.

Drury, Nevill. *Dictionary of Mysticism and the Occult*. New York: Harper and Row, 1985.

Drury, Nevill, and Stephen Skinner. *The Search for Abriaxas*. London: Neville Spearman, 1972.

Dudeney, Henry E. *536 Puzzles and Curious Problems*. New York: Charles Scribner's Sons, 1967.

Dudley, Underwood. *Numerology, or, What Pythagoras Wrought*. Washington, DC: Mathematical Association of America, 1977.

Duncan, Dewey. "Comment on Problem 84." *Mathematics Magazine* (Sept/Oct. 1966): 255.

———. "Heterosquares—Problem 84." *Mathematics Magazine* (Sept./Oct. 1965): 250.

Eberhard, Wolfram. *Chinese Symbols: Hidden Symbols in Chinese Life and Thought*. London: Routledge and Paul, 1986.

Eitel, E. J. *Feng Shui or the Rudiments of Natural Science in China*. Bristol: Pentacle Books, 1979. Reprint of 1873 Trubner edition.

Emanouilidis, Emanuel. "Powers of Magic Squares." *Journal of Recreational Mathematics* 29 (1998): 176–77.

Encyclopedia-Guide: China. Geneva: Nagel Publishers, 1984. 540–46.

Enge, Roger. "Reader Reflections: Magic-Square Designs." *Mathematics Teacher* 77 (November 1984): 596.

van den Essen, A. "Magic Squares and Linear Algebra." *American Mathematical Monthly* 97 (1990): 60–62.

Eves, Howard. *An Introduction to the History of Mathematics.* Philadelphia: Saunders Publishing, 1990.

Falkener, Edward. *Games Ancient and Oriental and How to Play Them.* New York: Dover Publications, 1961. Reprint of Longman and Green 1892 edition.

Ferguson, John. *An Illustrated Encyclopedia of Mysticism and the Mystery Religions.* London: Thames and Hudson, 1976.

Folkerts, Menso. "Zur Frühgeschichte der magischen Quadrate in Westeuropa." *Sudhoffs Archiv* 65 (1981): 313–38.

Forke, Alfred. *The World Conception of the Chinese.* New York: Arno Press, 1975.

Frawley, David. "The Five Elements, East and West." *Chinese Culture* 22 (1981): 57–64.

Franklyn, Julia. *A Survey of the Occult.* London: Arthur Barker, 1935.

Fults, John. *Magic Squares.* La Salle, IL: Open Court, 1974.

Fung Yu-lan. *A History of Chinese Philosophy.* 2 vols. Translated by Derk Bodde. Princeton: Princeton University Press, 1953.

Gardner, Martin. "Some New Discoveries about 3 × 3 Magic Squares." *Math Horizons* (February 1998): 11–13.

———. "The Magic of 3 x 3." *Quantum* 6 (1996): 24–28.

———. *Penrose Tiles to Trapdoor Ciphers.* New York: W. H. Freeman, 1989.

———. "The Combinatorial Basis of the 'I Ching', The Chinese Book of Divination and Wisdom." *Scientific American* 230 (1974): 108–13.

———. *The Numerology of Dr. Matrix.* New York: Simon and Schuster, 1967.

———. *The Scientific American Book of Mathematical Puzzles and Diversions.* New York: Simon and Schuster, 1960.

Gibson, Walter Brown. *The Magic Square Tells You Past, Present and Future.* New York: G. Sully, 1927.

Girardot, Norman. *Myth and Meaning in Early Daoism.* Berkeley: University of California Press, 1983.

Glaser, Anton. *History of Binary and Other Nondecimal Numeration.* Southampton, PA: Anton Glaser, 1971.

Godwin, David. *Cabalistic Encyclopedia.* St. Paul, MN: Llewellyn Publishers, 1979.

Golvers, Noël, ed. *The Astronomia Europaea of Ferdinand Verbiest S. J. (Dillingen, 1687): Text, Translation, Notes and Commentaries.* Nettetal: Steyler Verlag, 1993.

Goodman, Howard L. "Paper Obelisks: East Asia in the Vatican Vaults." In *Rome Reborn:The Vatican Library and Renaissance Culture*, edited by Anthony Grafton, 251–92. Washington, DC: Library of Congress, 1993.

Gordon, Antoinette K. *Tibetan Religious Art*. New York: Paragon Book Reprint Corp., 1963. Reprint of Columbia University edition, 1952.

Gordon, B. L. "Sacred Directions, Orientation, and the Top of the Map." *History of Religions* 10 (1971): 211–27.

Graham, A. C. *Yin-Yang and the Nature of Correlative Thinking*. Singapore: The Institute of East Asian Philosophy, 1986.

Granet, Marcel. *La Pensée chinoise*. Paris: Renaissance du livre, 1934.

———. *Chinese Civilization*. London: Kegan Paul, Trench, Trubner & Co., 1930.

de Gubernatis, Angelo. *La Mythologies des Plantes*. New York: Arno Press, 1978.

Guiley, Rosemary Ellen. *The Encyclopedia of Witches and Witchcraft*. New York: Facts on File, 1989.

Gurevich, Efim. *Toaina drevnego tatiamana*. Moscow: NAUKA, 1969.

Haq, Syed Nomaul. *Names, Natures and Things: The Alchemist Jabir ibn Hayyan and his Kitab al-Ahjar*. Boston: Kluwer Academic Publishers, 1994.

Hawkins, Gerald. *Stonehenge Decoded*. New York: Doubleday, 1965.

He Junshou. "Mingdai Beijing cheng jianzhu" [The Mathematical Base for Urban Planning of Beijing City in the Ming Dynasty]. *Guang Ming Ri Bao* (August 1986).

Heath, Royal Vale. *Mathemagic: Magic, Puzzles, and Games with Numbers*. New York: Dover Publications, 1953.

Henderson, John B. *The Development and Decline of Chinese Cosmology*. New York: Columbia University Press, 1984.

Hinnells, John R., ed. *The Facts on File Dictionary of Religions*. New York: Facts on File, 1984.

Ho Peng Yoke. "Chinese Magic Squares: Mathematics, Myth and Philosophy." *Kertas-kertasPersidangan Antarabangsa Pengajian Tionghoa* [Collected Papers: InternationalConference on Chinese Studies] Kuala Lumpur: November 20-21, 1993, 345–72.

———. *Li, Qi and Shu: An Introduction to Science and Civilization in China*. Hong Kong: Hong Kong University Press, 1985.

Ho Peng Yoke. "Magic Squares in East and West." *Papers on Far Eastern History* 8 (1973): 115–41.

Holmes, R. "The Magic Magic Square." *The Mathematical Gazette* (1970): 376.

Holroyd, Stuart. *Magic Words and Numbers*. New York: Doubleday and Company, 1976.

Hu, Wei. *I'tu ming-pien*. Taipei: Kuang-wen, 1971. Reprint of 1706 edition.

Hulse, David. *The Key of It All*. St. Paul, MN: Llewellyn Publications, 1994.

Hwa Suk Hahn. "Another Property of Magic Squares." *Math Horizons* (February 1998): 11–13.

Hwang, Ming-chong. "Ming-tang: Cosmology, Political Order and Monuments in Early China." Ph.D. dissertation, Department of East Asian Languages and Civilization, Harvard University, 1996.

Ifrah, Georges. *From One to Zero: A Universal History of Numbers*. New York: Viking Penguin, 1985.

Jones, William. *Credulities Past and Present*. London: Chatto and Windus, 1880.

Joseph, George Cheverghese. *The Crest of the Peacock: Non-European Roots of Mathematics*. London: I. B. Tauris, 1991.

Kapadia, Hiralal R. "A Note on Jaina Hymns and Magic Squares." *Indian Historical Quarterly* 10 (1934): 140–53.

Karpenko, Vladimír. "Two Thousand Years of Numerical Magic Squares." *Endeavour* 18 (1994): 147–53.

———. "Between Magic and Science: Numerical Magic Squares." *Ambix* 40 (November 1993): 121–28.

Katz, Victor J. *A History of Mathematics*. New York: HarperCollins, 1993.

Kenny, Margaret J. "An Art-Full Application Using Magic Squares." *Mathematics Teacher* 75 (1982): 83–89.

Klibansky, Raymond, Erwin Panofsky, and Fritz Saxl. *Saturn and Melancholy: Studies in the History of Natural Philosophy, Religion, and Art*. New York: Basic Books, 1964.

Koestler, Arthur. *The Sleepwalkers*. New York: Grosset & Dunlap, 1963.

Kramer, Samuel. *Mythologies of the Ancient World*. New York: Doubleday, 1961.

Kusuba, Takanori. "Combinatorics and Magic Squares in India: A Study of Narayana Pandita's Ganitakaumudi, Chapters 13-14." Ph.D. dissertation, Department of History, Brown University, May 1993.

Lach, Donald F. "Leibniz and China." *Journal of the History of Ideas* 6 (1945): 435–55.

Lagerwey, John. *Taoist Ritual in Chinese Society and History*. New York: Macmillan, 1987.

Lam Lay Yong. *A Critical Study of Yang Hui Suan Fa: A Thirteenth-Century Mathematica Treatise*. Singapore: Singapore University Press, 1977.

Lam Lay Yong and Ang Tian Se. *Fleeting Footsteps: Tracing the Conception of Arithmetic and Algebra in Ancient China*. Singapore: World Scientific, 1992.

Lawlor, Robert. *Sacred Geometry*. London: Thames and Hudson, 1982.

Legeza, Laszlo. *Tao Magic: The Secret Language of Diagrams and Calligraphy*. London: Thames & Hudson, 1975.

Legge, James. *The I-Ching*. New York: Dover Publications, 1963. Reprint of 1899 edition.

———. *The Chinese Classics*. 5 vols. Hong Kong: Hong Kong University Press, 1960. Reprint of 1865 edition.

Leibniz, Gottfried Wilhelm. *Writings on China*. Translated by Daniel J. Cook and Henry Rosemont. Chicago: Open Court, 1994.

———. *Discourse on the Natural Theology of the Chinese.* Translated by Henry Rosemont, Jr. and Daniel Cook. Honolulu: University of Hawaii Press, 1977.

Levi-Strass, Claude. *The Savage Mind.* Chicago: University of Chicago Press, 1966.

Levy, Raphael. *The Astrological Works of Abraham ibn Ezra.* Baltimore: Johns Hopkins University Press, 1927.

Levy, Raphael, and Francisco Cantera, eds. *The Beginnings of Wisdom: An Astrological Treatise by Abraham ibn Ezra.* Baltimore: Johns Hopkins Univeristy Press, 1939.

Li Yan and Du Shiran. *Chinese Mathematics: A Concise History.* Translated by John Crossley and Anthony Lun. Oxford: Clarendon Press, 1987.

Lindberg, David. *The Beginnings of Western Science.* Chicago: University of Chicago Press, 1992.

Lip, Evelyn. *Feng Shui: Environments of Power: A Study of Chinese Architecture.* London: Academic Editions, 1995.

———. *Chinese Numbers: Significance, Symbolism, and Tradition.* Singapore: Times Books International, 1992.

———. *Chinese Geomancy.* Singapore: Times Books International, 1979, 10.

Liu Da. *I-Ching Numerology.* San Francisco: Harper & Row Publishers, 1979.

Liu, Laurence G. *Chinese Architecture.* New York: Rizzoli International, 1989.

Liungman, Carl G. *Dictionary of Symbols.* Santa Barbara, CA: ABC-CLIO, 1991.

Loria, Gino. *Le Scienze esatte nell'Antica Grecia.* Milan: 1914.

Loewe, Michael. *Divination, Mythology and Monarchy in Han China.* Cambridge: Cambridge University Press, 1994.

de la Loubère, Simon. *A New Historical Relation of the Kingdom of Siam by Monsieur de la Loubère, Envoy-Extraordinary from the French King to the King of Siam in the years 1687 and 1688.* Translated by A. P. Gent. London: Horne Saunders & Bennet, 1693; from the French edition, Paris, 1691; reprinted by John Villieis, ed., Bangkok: White Lotus, 1986.

Loewenstein, P. J. "Swastika and Yin-Yang." *China Society Occasional Papers.* London: The China Society, 1942.

Lyon, Betty Clayton. "Using Magic Borders to Generate Magic Squares." *Mathematics Teacher* 77 (1984): 223–26.

MacMahon, P. A. "Magic Squares and Other Problems on a Chessboard." *Proceedings of the Royal Institute of Great Britain* 17 (February 1892): 50–61.

Major, John S. "Myth, Cosmology, and the Origins of Chinese Science." *Journal of Chinese Philosophy* 5 (1978): 3–20.

———. "The Five Phases, Magic Squares and Schematic Cosmology." *Journal of the American Academy of Religious Studies* 50 (1976): 133–66.

Marder, Clarence. *The Magic Squares of Benjamin Franklin.* New York: The Brick Row Book Shop, 1940.

Marsh, Benjamin. *Analysis of the Properties of Numbers as Discovered and Applied to the Construction of Magic Squares.* Helena, MT: C. K. Wells Co., 1892.

Martzloff, Jean Claude. *A History of Chinese Mathematics*. New York: Springer Verlag, 1997.

Maspero, Henri. "Le Ming-Thang et la Crise Religieuse Chinoise avant les han." *Mélanges Chinois et Bouddhiques* 9 (1951): 1–71.

Mayers, William Frederick. *The Chinese Reader's Manual*. Shanghai: American Presbyterian Press, 1910.

McCoy, John Calvin. "Manuel Moschopoulos's Treatise on Magic Squares." *Scripta Mathematica* 8 (1941): 15–26.

McGregor, Richard E. "The Maxwell Davies Sketch Material in the British Library." *Tempo* (April 1996): 9–19.

Meller, Walter Clifford. *Old Times: Relics, Talismans, Forgotten Customs and Beliefs of the Past*. London: T. Werner Laurie, 1925.

Michelle, John. *The Dimensions of Paradise: The Proportions and Symbolic Numbers of Ancient Cosmology*. New York: Harper & Row, 1988.

———. *The View Over Atlantis*. New York: Ballantine Books, 1969.

Miller, Laurel. *Kabbalistic Numerology or the True Science of Numbers*. New York: Metaphysical Publishing Co., 1921.

Moran, Jim. *The Wonders of Magic Squares*. New York: Random House, 1982.

Morgan, Carole. "Les 'Neuf Palais' dans les Manuscrits de Touen-Houang." *Nouvelles Contributions aux Etudes de Touen-Houang*. Geneva: Librairie Droz, 1981.

Morgan, Harry T. *Chinese Symbols and Superstitions*. South Pasadena, CA: Ione Perkins, 1942.

Motta, Marcelo Ramos. *The Art and Practice of Caballa Magic*. New York: Samuel Weiser, 1977.

Mungello, David E. *Curious Land: Jesuit Accommodation and the Origins of Sinology*. Stuttgart: Franz Steiner Verlag, 1985.

———. *Leibniz and Confucianism: The Search for Accord*. Honolulu: University of Hawaii Press, 1977.

Murdoch, John. "Transmission and Figuration: An Aspect of the Islamic Contribution to Mathematics, Science and Natural Philosophy in the Latin West." *Proceedings of the First Symposium for the History of Arabic Science, University of Aleppo, April 5–12, 1976*. Edited by Ahmad Y. al-Hassan et al., 108–22.

Nakamura, Hajime. *Ways of Thinking of Eastern Peoples: India, China, Tibet and Japan*. Honolulu: East-West Center Press, 1964.

Nasr, Seyyed Hossein. *An Introduction to Islamic Cosmological Doctrines*. Albany, NY: SUNY Press, 1993.

———. *Science and Civilization in Islam*. Cambridge, MA: Harvard University Press, 1968.

Nauert, Charles G. *Agrippa and the Crisis of Renaissance Thought*. Urbana, IL: University of Illinois Press, 1965.

Needham, Joseph. *Science and Civilisation in China.* Vol. 2, *History of Scientific Thought,* 1956; Vol. 3, *Mathematics and the Sciences of the Heavens and the Earth,* 1959; Vol. 5, *Chemistry and Chemical Technology,* 1959. Cambridge: Cambridge University Press.

Neihardt, John. *Black Elk Speaks.* Lincoln, NE: University of Nebraska Press, 1972.

Neugebauer, O. *The Exact Sciences in Antiquity.* New York: Dover Publications, 1969. 168–69.

Nowotny, Karl Anton. "The Construction of Certain Seals and Characters in the Work of Agrippa of Nettesheim." *Journal of Warburg and Courtauld Institute* 12 (1949): 46–57.

Pagni, David. "Magic Squares: Would You Believe. . . ?" *Arithmetic Teacher* 21 (1974): 439–41.

Palter, Robert, ed. *Towards Modern Science.* 2 vols. New York: Noonday Press, 1961.

Pang, Pu. "Origins of the Yin-Yang and Five Elements Concept." *Social Sciences in China* (Spring 1985): 91–131.

Panofsky, Edwin. *Albrecht Dürer.* Princeton, NJ: Princeton University Press, 1943.

Pingree, David, ed. *Picatrix: The Latin Version of the Ghayat al-Hakim.* London: The Warburg Institute of London, 1986.

Pinzka, Charles. "Heterosquares." Problem 84. *Mathematics Magazine* (September/October 1965): 250–52.

Pullan, J. M. *The History of the Abacus.* New York: Frederick A. Praeger, 1969.

Reding, Jean-Paul. "Greek and Chinese Categories: A Reexamination of the Problem of Linguistic Relativism." *Philosophy East and West* 36 (1986): 349–74.

Roberts, David. "Peter Maxwell Davies: *Ave Maris Stella.*" *Contact* 19 (1978): 26–31.

Robertson, John P. "Magic Squares of Squares." *Mathematics Magazine* 69 (1996): 289–93.

Ronan, C. E., and B. B. Oh, eds. *East Meets West: The Jesuits in China (1582–1773).* Chicago: Loyola University Press, 1988.

Rose, David M. "Magic Squares and Matrices." *Mathematical Gazette* (1973): 36–39.

Rosemont, Jr., Henry. *Explorations in Early Chinese Cosmology.* Chico, CA: Scholars Press, 1976.

Rosu, Arion. "Les carrés magiques indiens et l'histoire des idées en Asie." *Zeitschrift der Deutschen Morgan* 139 (1989): 120–58.

―――. "Etudes ayurvédiques III 'Les carrés magique dans la médecine indiénne.'" In *Studies in Indian Medical History,* edited by Dominik Wujastyk and Jan Meulenbeld Groningen: Egbert Forsten, 1987.

Rouach, David. *Les Talismans: Magic et tradition juives*. Paris: Albin Michel, 1989.
Sallows, Lee. "Alphamagic Squares." *Abacus* 4 (1986): 28–45; 4 (1987):20–29, 43.
Saso, Michael. "What Is the Ho-T'u?" *History of Religions* 17 (1978): 399–416.
———. *Taosim and the Rite of Cosmic Renewal*. Seattle: Washington State University Press, 1972.
Schaefer, Edward H., ed. *Ancient China*. New York: Time-Life Books, 1967.
Schimmel, Annemarie. *The Mystery of Numbers*. New York: Oxford University Press, 1993. Revised version of *Mysterium der Zahl* by Franz Enders. Munich: Diederich Verlag, 1984.
———. *Mystical Dimensions of Islam*. Chapel Hill: The University of North Carolina Press, 1975.
Schrire, T. *Hebrew Amulets: Their Decipherment and Interpretation*. London: Routledge, Kegan Paul, 1966.
Schubert, Hermann. *Mathematical Essays and Recreations*. Chicago: Open Court, 1903.
Schuon, Frithjof. *Dimensions of Islam*. London: George Allen and Unwin, 1970.
Seidenberg, Abraham. "The Ritual Origin of Geometry." *Archive for History of Exact Sciences* 1 (1962): 488–527.
Seligmann, Kurt. *Magic, Supernaturalism and Religion*. New York: Grosset & Dunlap, 1968.
———. *The Mirror of Magic*. New York: Pantheon Books, 1948.
Selin, Helaine. *Encyclopaedia of the History of Science, Technology and Medicine in Non Western Cultures*. Boston: Kluwer Academic Publishers, 1997.
Sherrill, W. A., and W. K. Chu. *An Anthology of I-Ching*. London: Routledge & Kegan Paul, 1977.
Shortreede, Captain. "On an Ancient Indian Magic Square, Cut in a Temple at Gwalior." *Journal of the Asiatic Society of Bengal* 11 (1842): 292.
Shumaker, Wayne. *The Occult Sciences in the Renaissance: A Study of Intellectual Patterns*. University of California Press, 1972.
Singmaster, David, and John Hadley. "Problems to Sharpen the Young." *The Mathematical Gazette* (March, 1992): 102–26.
Skeat, Walter. *Malay Magic*. New York: Dover Publications, 1967.
Small, C. "Magic Squares over Fields." *American Mathematical Monthly* 95 (1988): 621–25.
Smith, David Eugene. *History of Mathematics*. New York: Dover Publications, 1958.
Smith, Richard J. *Chinese Almanacs*. Hong Kong: University of Hong Kong Press, 1992.
———. *Fortune-tellers and Philosophers: Divination in Traditional China*. Boulder, CO: Westview Press, 1991.
Smith, Richard J., and D. W. Y. Kwok, eds. *Cosmology, Ontology and Human Efficacy: Essays in Chinese Thought*. Honolulu: University of Hawaii Press,

1993.

Sofar, Anna. *The Sun Dagger*. Bethesda, MD: Atlas Video, 1993. Video recording.

Solomon, Bernard. "One is No Number in China and the West." *Harvard Journal of Asiatic Studies* 17 (1954): 253–60.

Soothill, William Edward. *The Hall of Light: A Study of Early Chinese Kingship*. New York: Philosophical Library, 1952.

Soror, A. L. *Western Mandalas of Transformation: Magic Squares, Tattwas, Qabalistic Talismans*. St. Paul, MN: Llwellyn Publications, 1995.

Spector, Sheila. *Jewish Mysticism: An Annotated Bibliography on the Kabbalah in English*. New York: Garland Press, 1984.

Spence, Jonathan. *The Memory Palace of Matteo Ricci*. New York: Viking Penguin, 1984.

Spitz, Lewis. "Reuchlin's Philosophy: Pythagoras and Cabala for Christ." *Achiv für Reformationsgeschichte* 47 (1956): 1–21.

Stapleton, H. E. "Ancient and Modern Aspects of Pythagoreanism." *Osiris* 13 (1958): 12–53.

———. "The Gnomon." *Ambix* 6 (1957): 1–9.

———. "The Antiquity of Alchemy." *Ambix* 5 (1953): 1–43.

Strang, Gilbert. *Linear Algebra andIts Applications*. New York: Academic Press, 1976.

Stroyls, John. "Survey of the Arab Contribution to the Theory of Numbers." *Proceedings of the First International Symposium for the History of Arabic Sciences*, vol. 2. Aleppo: Institute for the History of Arabic Sciences, April 5-12, 1976. 168–79.

Sung, Z. D. *The Symbols of Yi King; or, The Symbols of the Chinese Logic of Change*. New York: Paragon Book Reprint Corp., 1969. Reprint of 1934 Shanghai edition.

Swetz, Frank. "If the Squares Don't Get You—The Circles Will." *Mathematics Teacher* 73 (1980): 67–72.

———. *Mathematics Education in China: Its Growth and Development*. Cambridge, MA: MIT Press, 1974.

———. "Mysticism and Magic in the Number Squares of Old China." *Mathematics Teacher* 71 (1978): 50–56.

Swetz, Frank J., ed. *From Five Fingers to Infinity: A Journey through the History of Mathematics*. Chicago: Open Court, 1994.

Swetz, Frank, et al., eds. *Learn from the Masters*. Washington, DC: Mathematical Association of America, 1995.

Thite, G. U. *Medicine: Its Magico-Religious Aspects According to the Vedic and Later Literature*. Poona: Continental Prakashan, 1982.

Thompson, A. C. "Odd Magic Powers." *American Mathematical Monthly* 101 (1994): 339–42.

Too, Lillian. *Chinese Numerology in Feng Shui*. Kuala Lumpur: Konsep Books, 1994.

Trigg, Charles W. "Comment on Problem 84." *Mathematics Magazine*

(September/October 1965): 250–52.

———. "Magic Square as a Determinant." *American Mathematical Monthly* 61 (1949): 33–37.

———. "The Mathematician and the Jester." *American Mathematical Monthly* 55 (September 1948): 429–30.

Tyrey, Bradford, and Marcus Brinkman. "The *Luo-shu* as *Taiji* Boxing's Secret Inner-Sanctum Training Method." *Journal of Asian Martial Arts* 5 (1996): 75–79.

Tyson, Donald. *Three Books of Occult Philosophy Written by Henry Cornelius Agrippa*. St. Paul, MN: Llewellyn Publications, 1995.

Vicker, Brian, ed. *Occult and Scientific Mentalities in the Renaissance*. Cambridge: Cambridge University Press, 1984.

Vijayaraghavan, T. "On Jaina Magic Squares." *The Mathematics Student* 9 (1941): 97–102.

van der Waerden, B. L. *Science Awakening*. New York: Oxford University Press, 1961.

Vogel, Kurt. *Die Practica des Algorismus Ratisbonensis*. Munich: C. H. Beck, 1954.

Walker, G. W. "The Mathematician and the Jester—Problem E791." *American Mathematical Monthly* 53 (1948): 429–30.

Wong, Ann-Lee. "Hollow Magic Squares." *Mathematics in School* (March 1995): 23–25.

Wang, Zhongshu. *Han Civilization*. New Haven, CN: Yale University Press, 1982.

Ward, James. "Vector Spaces of Magic Squares." *Mathematics Magazine* 53 (1980): 108–11.

Webster, Richard. *Talisman Magic: Yantra Squares for Tantric Divination*. St. Paul, MN: Llewellyn Publications, 1995.

Werner, E. J. *A Dictionary of Chinese Mythology*. New York: Julian Press International Publishers, 1961.

Westcott, W. Wynn. *Numbers: Their Occult Power and Mystic Virtue*. London Theosophical Society, 1974.

Westermarck, Edward. *Ritual and Belief in Morocco*. 2 vols. London: Macmillan & Co., 1926.

Wheatley, David. *The Devil and All His Works*. New York: American Heritage Press, 1971.

Wilhelm, Richard. *'I Ging'; Das Buch der Wandlungen*. Jena: Diederichs, 1924.

Willcox, William B., ed. *The Papers of Benjamin Franklin*. Vol. 15, 171–72. New Haven, CN: Yale University Press.

Williams, S. Wells. *The Middle Kingdom*. 2 vols. New York: Charles Scribner's Sons, 1883.

Wittfogel, Karl. *Oriental Despotism*. New Haven, CN: Yale University Press, 1957.

———. "Die Theorie der orientalishen Gesellschaft." *Zeitschrift fur*

Sozialforschung 7 (1938): 90.

Wong, Eva. *Feng-shui: The Ancient Wisdom of Harmonious Living for Modern Times*. Boston: Shambhala Publications, 1996.

Wright, Arthur F. "The Cosmology of the Chinese City." *The City in Late Imperial China*. Edited by G. William Skinner. Stanford, CA: Stanford University Press, 1977. 33–75.

Yang, Hsiung. *The Canon of Supreme Mystery*. Translated by Michael Nylan. Albany, NY: State University of New York Press, 1993.

Yates, Frances A. *The Occult Philosophy in the Elizabethan Age*. London: Routledge, Kegan Paul, 1979.

Yoon, Hong-key. *Geomantic Relations between Culture and Nature in Korea*. Taipei: Chinese Association for Folklore, 1976.

Zaslavsky, Claudia. *Africa Counts: Number and Pattern in African Culture*. Boston: Prindle, Weber & Schmidt, 1973.

Illustrations Acknowledgments

1.2 *Biblioteca Apostolica Vaticana:* Borg. cin 397, fol. 142. Plate 213, *Rome Reborn: The Vatican Library and Renaissance Culture*, ed. Anthony Grafton (Washington: Library of Congress, 1993), p. 288.

2.1 From Michael Saso, *Taoism and the Rite of Cosmic Renewal* (Seattle: Washington State University Press, 1972), p. 119. Reproduced from Lai Chih-te, *Lai Chu I-ching T'u Shuo*. With permission.

2.2 Hu Wei, *I-t'u ming-pien* (1706) (Taipei: Kuang-wen reprint, 1971) 5:9.

3.2(a) From James Legge, *The Chinese Classics*, 5 vols. (Hong Kong: Hong Kong University Press, 1960) [Reprint of 1865 edition] III: XIII. With permission.

3.9 Ho Peng Yoke, *Li, Qi and Shu: An Introduction to Science and Civilization in China* (Hong Kong: Hong Kong University Press, 1985), p. 20. With permission.

4.1 Liu Dunzhen, *Zhongguo gudai jianzhu shi*. Beijing, 1980, fig. 30-4.

4.3 Hu Wei, *I-t'u ming-pien* (Taipei: Kuang-wen reprint, 1971) 2:106.

4.5 As reproduced in Ho Peng Yoke, "Chinese Magic Squares: Mathematics, Myth and Philosophy." *Collected Papers: International Conference on Chinese Studies* (Kuala Lumpur: University of Malaya, Nov. 20-21, 1993), p. 364.

4.7 Ho Peng Yoke, *Li, Qi and Shu: An Introduction to Science and Civilization in China* (Hong Kong: Hong Kong University Press, 1985), p. 35. With permission.

4.14 As reproduced in Lars Berglund, *The Secret of Luo Shu: Numerology in Chinese Art and Architecture* (Lund: Lund University, 1990).

6.10   As reproduced in Antoinette K. Gordon, *Tibetan Religious Art* (New York: Paragon Book Reprint, 1963), p. 89.

6.19   W. Ahrens and Alfred Maass, *Etwas von magischen Quadraten in Sumatra und Celebes* (Berlin, 1916), pp. 251, 253.

6.24(a) Georges Ifrah, *The Universal History of Numbers: From Prehistory to the Invention of the Computer* (New York: Wiley Publishers, 2000), p. 262. Reprinted by permission of John Wiley & Sons, Inc.

6.32   From Girolamo Cardano, *Practice arithmetice et mensurandi singularis* (Milan, 1539).

7.17   Reproduced from C. F. Sallows, "Alphamagic Squares: Adventures with Turtle Shell and Yew Between the Lowlands of Logology." *Abacus* (Fall, 1986), pp. 28–45, p. 31. With permission of Springer-Verlag, New York.

7.26   Reproduced from Claude Bragdon, *The Frozen Fountain* (New York: Books for Libraries Press, 1970), pp. 44, 76. Copyright 1932 and renewed 1960 by Henry Bragdon. Reprinted by permission of Alfred A. Knopf, a Division of Random House.

7.27   Reproduced from Claude Bragdon, *The Frozen Fountain* (New York: Books for Libraries Press, 1970), p. 76. Copyright 1932 and renewed 1960 by Henry Bragdon. Reprinted by permission of Alfred A. Knopf, a Division of Random House.

8.5   Figure 3, "The Aztex five world regions," by Raymond Turver. Reproduced from *The Penguin Dictionary of Religions*, ed. John R. Hinnels (Penguin Books, 1984), p. 95 With permission of Penguin Books, U.K.

Index

abacus, 22
Abbé Poignard, 118
   Traité des Quarrés sublimes,
     118
abjad, 94, 95, 100, 104
Abraham ben Meir ibn Ezra
   (Abenezra), 107
Abu'l-Abbas al-Buni, 97, 107
   Shams al-Ma'arif al-Kuba, 97
Abu Ma'shar, 107
Agrippa von Nettesheim, Henricus
   Cornelius, 114, 116, 154
   De occulta philosophia,
     114, 116
alchemy, 30, 98, 145
   "elixirs of Immortality," 145
Alcuin of York, 109
   Propositiones ad acuendos
     juvenes, 109
Alfonso X, 107
Al-Ghazali of Tus, 97
Al-Majriti, 107
   Ghajat al-Hakim, 107
almanac, 89
alphamagic square, 133
amulet, 101, 102, 107
   planetary, 105
ancient theology, 7
Ando, Yueki, 90, 91
   Kigu hosu, 90

antimagic square, 130
   heterosquare, 131
Apollonius of Tyana, 93, 152
   Sirr al-Khaliqa, 93
arithmology, 80
astrology, 56, 98

Bachet de Meziriac, Claude-Gaspar,
   118
   Problemes plaisants et
     delectables, 118
bagua. See Eight Trigrams
Bao Qishou, 77
BEDUH, 101
Beidou [the Plough constellation], 50,
   56
Berglund, Lars, xiii
Book of Balances, 94
Book of Changes. See *Yijing*
Book of Rites, 12, 44, 45
Bouvet, Joachim, 6
Boyang, Fu, 30
Bragdon, Claude, 141
Brethren of Purity. See Ikhwan
   as-Safa
Browne, C. A., 135
   magic squares, 136, 144
Buddhism, 82, 90
"burning of the books," 12

Cai, Yong, 41
Cai, Yuanding, 16
Cammann, Schuyler, xii, 66, 70, 84, 87, 154
Cardano, Girolamo, 116
   Practica arithmetice, 116
Chaos theory, 164
Chen Dawei, 77, 90
   Suanfu tongzong, 90
Chen Weiming, 147
Chinese mathematics education, xii
   Classic of the Nine Halls of the Yellow Emperor, 14
Confucius, 13
   Confucianism, 13
"conquest cycle," 33, 34
cosmic harmony, 19
cosmic mirror, 62
cosmogram, 157
Cosmos of Alexandria, 158
   Topographia Christiana, 158
Counting rods, 154. *See also* rod numerals
court jester problem, 129
crystalline spheres, 5, 159
   "music of the spheres," 162
cycle of reincarnation, 88

Dagomari dell'Abaco, Paolo, 109
   Trattato d'Abbaco, 109
Dai, De 47
   Dadai lyi, 47
Dao, 27
   Daoism, 13, 19, 30
Daozang, 53
Davies, Peter Maxwell, 140
   Ave Maris Stella, 140
Dee, John, 116
De Ferro, Scipione, 110
de la Hire, Phillippe, 8, 108, 118
   "Method of la Hire," 118
de la Loubère, Simon, xi, 118, 119
   Du Royaume de Siam, 118
divination, 56, 83, 89
Dong, Zhongshu, 34

d'Ons-le-Bray, 119
Dragon Horse, 10
Duncan, Dewey, 131
Dürer, Albrecht, 109, 110
   Melancholia I, 109, 110

"Earth square," 58
Eight Trigrams, 53
   characteristics, 55
   Earlier Heaven Circle, 54
   Later Heaven Circle, 54
electronic computer, 143, 159
Emperor Yu, 10, 21, 50
Enders, Frans, 80
Enuma Elish, 79

Facing Star, 58
Fang, Zhongtong, 77
fengshui, 2, 39, 58
Fermat's Last Theorem, 132
figures of Fohi (Fuxi), 7
five, 21, 25, 26, 37
   agents, 31
   as auspicious, 22
   colors, 31
   directions, 36
   elements, 32
   as marriage number, 81
   processes, 31
   punishments, 22
   sacred mountains, 29
   sounds, 31
   weapons, 22
Fludd, Robert, 116
Forbidden City, 64
fortune-telling, 59–61, 103
four
   angels, 101
   corners of the Earth, 159, 163
   elements, 32, 81, 93, 94, 95, 159
   human traits, 159
   humors, 109, 159
   principal rivers, 29
   seasons, 31, 159

Franklin, Benjamin, 119
Frenicle de Bessy, Bernard, 119
  Traité des Triangles Rectangle en nombres, 119
Frost, A. H., 85, 119
Fulani, 104
Fuxi, 9, 54

Galilei, Galileo, 160
games, 104
Gardner, Martin, 126, 132, 157
Garga, 84
  Gargasamhita, 84, 85
gematria, 80
Giorgi, Francesco, 116
Golvers, Noël, 3
Granet, Marcel, xii, 156, 159
"Great Calendar Stone," 157
Grimaldi, Claudio, 6

Hahn, Hwa Suk, 126
Harvanian beliefs, 105, 106
hetu, 10, 16, 36, 37, 54, 155
hexagrams, 39, 42, 53
Ho, Peng Yoke, XIII, 34
Holmes, R., 125
Huainanzi, 27
hydraulic society, 12

Ibn al-Lubudi, 97
Ibn al-Samh, 107
Book of the Plates of The Seven Planets, 107
Ibn Sina (Avicenna), 97, 100
Ikhwan as-Safa, 95
  Rasa'il, 95, 96
Ilm al-asrar, 104
Ishtar, 80
Islam, 93
isopsephia, 80, 82, 84

Jabir ibn Hayyan (Geber), 93

Kitab al-Manwazin al-Saghir, 93
Jesuit missionaries, 3, 5, 32
Jiu, Gong. *See* Nine Halls

Kabbala, 80, 108, 110, 114
  of the Nine Chambers, 112
Kangxi Emperor, 7
Kaogong ji, 63
Kepler, Johann, 159
King Mi, 134
"King's Assemblage," 47
kowtow, 21

Lane, Edward, 104
Legge, Edward, 156
Leibniz, Gottfried Wilhelm, 6, 7, 8, 118
  binary arithmetic, 7
Libros del Saber de Astrologia, 107
Liu, Xiang, 63
Li shu, 134
Lloyd, Sam, 130
Logan, James, 119
Luo River writing, xi, 2
luoshu, xi, 10, 14, 34, 54, 55, 80
  anthropomorphically described, 14
  basis of mathematics, 23
  charm, 61
  definition of, 2
  kabbala: *Aiq Beker,* 112
  Mingtang association, 47
  uniqueness of, 125
Luxuriant Dews of the Springs and Autumns, 36

magic circles, 76, 77, 86
magic cube, 8, 133
magic square, xi
  augmented, 70
  border, 103
  composite, 75
  definition of, 2

division, 132
doubly-even, 119
hollow, 130
higher order, 77, 90
normal, 124, 135
multiplicaton, 132
of order five, 69, 98
of order four, 68, 69, 83, 85, 87, 110
of order nine, 74–76
of order one, 121
of order seven, 73, 74, 96
of order six, 71–73
of order three, 84, 98, 108, 124
of order two, 122, 123
Magic Squares and Circles, 135
mandala, 154, 157, 160
Mandarin Squares, xi
mantra, 85, 86
"mansions of the moon," 94
medicine, 98
Minamoto, Tamenori, 90
  Kuchi-zusami, 90
Mingtang, 14, 23, 39, 40–48, 63, 151
Mohism, 13
Moschopoulos, Manual, 108
Muhammad, 100
Muhammad ibn Muhammad, 104

Nagarjuna, 84
Narayana, 86, 87
  Ganitakaumudi, 86, 87
Needham, Joseph, 154
Nelson, Harry, 132
nine, 22, 23
  calculations, 25
  categories, 96
  classes of merit, 21
  cycles, 21
  huo, 13
  marshes, 21
  Palaces, 9, 14, 89
  provinces, 2, 3
  regions, 40
  rivers, 21, 22

Nine Chapters on the Mathematical Art, 90
nine halls, 9, 16, 17, 50
  calculation, 14
north star, 44
numbers
  even, 80
  female, 163
  lucky, 79, 82, 163
  male, 163
  odd, 80
  triangular, 81
  unlucky, 79, 82, 163
numerology, 20, 64, 81, 114

oracle-bone inscriptions, 20
ornamental designs, 141
Origin of Tree Worship, 134
Osage, 162

Paciolo, Luca, 109, 110
  De viribus quantitatis, 109
palm-reading, 92
Paracelsus, 116
  Archidoxa Magica, 116
Picatrix, 107
Plato, 135
  Timaeus, 135
Plutarch, 135
Posidonius, 93
proto-scientific theories, 160
Pythagoras of Samos, 80, 163
  reborn, 114
Pythagorean(s) 3, 95, 153
  numbers, 135
  world view, 154

Qi, 31, 147
Qin, Shihuangdi, 12

Rechnung auff der Linien und Federn, 117
Reuchlin, Johannes (Capnion), 114

De arte cabalistica, 114
Ricci, Matteo, 6
Riese, Adam, 116
Robertson, John 126, 127
   "Robertson squares," 126
rod numerals, 20

Sallows, Lee, 133, 134
Saturn, 107, 115
Sator square 152, 153
Sauveur, Joseph, 118, 119
Schubert, Hermann, 157
semi-magic square, 128
seven
   diseases, 80
   planetary squares, 109, 116, 117
   seals, 80
   spirits, 80
   winds, 80
shaman, 161
Shem Hameforash, 111
Shujing, 23
Shushu jiyi, 13
Singmaster, David, 108
Sioux, 2
Skeat, Walter, 102
Socrates, 138, 139
   Republic, 139
Soothill, William, 47
"square deal," 164
"Square of the Beast," 132
Stifel, Michael, 117
   Arithmetica integra, 118
   Stifelsche Quadrate, 118
Stonehenge, 162
Sun Dagger, 162
Sun Lutang, 147
swastika, 35

t'ai chi. *See taijiquan*
taijitu, 156
taijiquan, 30, 145–49
Taiyi [Sky Emperor] 48, 54
Takebe, Katohiro, 91

talisman, 86, 102, 105, 115
Temple of Heaven, 40
Temurah, 112
tetractys, 81, 82, 96, 135, 154
Tetragrammation, 111
Thabit ibn Qurra, 97
Thakkakra, Pheru, 86
   Ganitasara, 86
Theon of Smyrna, 122, 152
   Expositio, 122
Theodorus of Asine, 152
Three Sage Kings, 9
tortoise, 10, 12
torus, 99
Trigg, Charles W., 140
"Twelve Palaces," 42, 44

Varahamihira, 83
   Brhatsamhita, 83
Varignon, Pierre, 118
vector space, 127, 128
Verbiest, Ferdinand, 1, 3, 7
   Astronomia Europaea, 1
Vogel, Kurt, 108
Vrnda, 85
   Saddhayoga, 85

Wang, Mang, 49
Webster, Richard, 121
   Talisman Magic, 121
Westermarck, Edward, 103
"Wheel of the Universe," 157
Wudi, 48
Wuxing, 19, 31, 63, 90

Xu, Yue, 13
Xugu zhaiqi suanfu, 16

yang, 53, 143
   numbers, 5
Yang, Di, 13
Yang, Hui, 16, 65, 68

Yangtze, 12
yantra, 85, 86
Yellow River, 10, 12
yinyang, 27–31, 54, 63, 90, 155
  directions, 29
Yijing [Book of Changes], 7, 10, 13, 42, 54
yin, 53, 143
  numbers, 5
Yongle Emperor, 64

yubu, 39, 50, 51, 56, 58, 115, 136, 138, 140, 141, 155

Zhang, Chao, 77
Zhang, Sanfeng, 146, 147
Zhen, Luan, 14, 90
Zheng, Xuan, 14, 28
Zhuang, Zi, 9, 13
Zou, Yan, 31, 33, 40